Teaching GCSE Craft, Design and Technology

D1823495

This book is dedicated to my wife Vivien, without whose patience, loyal support and understanding it would not have appeared.

Teaching GCSE Craft, Design and Technology

David Rees

HODDER AND STOUGHTON
LONDON SYDNEY AUCKLAND TORONTO

ISBN 0 340 40877 4

First published 1987

Copyright © 1987 David Rees

Typeset by Tradespools Ltd, Frome, Somerset
Printed and bound in Great Britain for
Hodder and Stoughton Educational,
a division of Hodder and Stoughton Ltd,
Mill Road, Dunton Green, Sevenoaks, Kent,
by Butler & Tanner Ltd, Frome and London

Contents

Acknowledgments

Grateful acknowledgment is made for assistance and support given by the following: Mr Martin Trevor, Avon Local Education Authority Adviser for Craft, Design and Technology for permission to visit schools; Mr Michael Smith, Headmaster, Filton High School, and Mr Brian Bassett, Head of Craft, Design and Technology, Filton High School, for allowing photographs to be taken of pupils at work; Mr Trevor Taylor, Avon Teacher Adviser in Technology, for advice and encouragement and permission to use photographs of pupils at work in the Avon Schools Technology bus; Mr Brian Williams, Head of Craft, Design and Technology at Brimsham Green School, for his contribution to the schemes of work; Mr Kevin Dengate, Head of Craft, Design and Technology, the Graham School, Scarborough, also for his contribution to the section on schemes of work; the Secondary Examinations Council for permission to quote from their publications; London and East Anglian Group, Southern Examination Group, Northern Ireland Examinations Council, Welsh Joint Education Committee, Midland Examination Group, Northern Examining Association for permission to quote from their syllabuses and instructions for teachers; and to the many pupils who assisted with the photographs and provided work to be used to illustrate the text; British Thornton for supplying photographs of pupils using their equipment; Economatics (Education) Ltd for all their help; Edding (UK) Ltd for supplying sample materials.

My grateful thanks go to Mrs Joan Eede who contributed to the major share of typing the manuscript and for tolerating my poor handwriting.

I would also like to make a special acknowledgment to Mr Michael Rose, Chief Examiner for GCSE in Craft, Design and Technology, for his contribution to Chapter 2.

Introduction

Communication

The time of two subjects – Technical Drawing or Graphic Communication, and Craft, Wood or Metal Work taking place, often under the same roof, and being taught as unrelated disciplines, is over. There has never been such an important move in the history of craft education as the bringing together of the two disciplines – drawing and designing. For many of us these are inseparable and, with the emphasis in the practical subjects now being replaced by the activity of solving problems, the one cannot do without the other.

While developing ideas through drawing has now become an essential and vital ingredient of following a design process, the values of communication have not totally succumbed to the needs of the students involved with a practical outcome. Instead it is still recognised that there is a need to specialise in communication skills, and a rightful place has been given under the umbrella of 'Craft, Design and Technology' along with 'Design and Realisation' and 'Technology'.

The strand to be known as Design and Communication allows the opportunity for the emphasis to be placed upon developing illustrative techniques, but in common with the two other strands, problem-solving is a prime activity.

Problem-solving

Problem-solving is a relatively new concept in the history of craft education. Its arrival in the late 1960s has brought about considerable changes in attitudes and educational thinking amongst teachers and in the world of commerce and industry. The ability to identify problems, to bring about a change and to evaluate its outcome are important attributes to acquire at all levels. It is arguable that it is just as important to be able to resolve a minor domestic problem as it is to resolve a highly technological one.

Equal Opportunities

With the introduction of the Equal Opportunities Act the craft teacher found himself – not yet herself – confronted not only with the problems of coping with notions of 'designing' and 'problem-solving' but also with dealing with girls in what has been essentially a male domain for nearly a century. The new situation, if nothing else, has prompted the traditionalist to look more closely at his own teaching and to develop ideas that are more appropriate to the needs of his pupils, both boys and girls.

Future Implications

To keep up with the changes that have and are taking place, much modification will need to be made in the school workshops and drawing offices. With designing being an integral activity for the pupils of today and the future, the selection of the material from which to make the chosen design could be from any of the three main resistant materials that are already in common use, i.e. wood, metal and plastics. The choice of material being determined by the solution to a problem now brings about the need for multi-media workshops. Though very strong arguments have been made to retain a Woodwork room and a Metalwork room etc., this is no longer realistic, and it is important for the tools and materials to be readily available to allow the transition from an idea to artefact to be smoothly achieved. Similarly, the drawing office will now need to include worktops and the associated tools to enable models to be made. Such, then, are some of the changes which may be expected if they have not already taken place.

The Aim of This Book

As CDT teachers, we have been going through a period of considerable change. However, it is hoped that the reader will realise on reading this book that a lot of the good qualities which have evolved over the years in Craft Education will be preserved. There is still an important place for good craftsmanship, there is still a place for pride in one's work, there is still a need to produce drawings of quality. These virtues are integral to the success of CDT, no matter what other changes have taken place. So the teacher who has devoted much energy to the furtherance of these skills in his pupils must continue to do so, and steadily adapt to some of the newer notions.

It is also hoped that the teacher will be reassured of his or her competence to cope with implementation of CDT, not only as a teacher but also as an assessor. Where possible, the aspects of the old and new approaches of assessment have been identified in CSE and GCE to enable the transition from one system to another to be as smooth as possible.

1 GCSE Assessment Criteria and Examination Techniques

NATIONAL CRITERIA

The National Criteria are little more than a set of rules with which every subject must comply. They deal with the structure content of the subjects and the administrative requirements. A body known as the Secondary Examinations Council is a consultative committee whose responsibility it is to see that the National Criteria are implemented, reviewed when necessary and to advise the Secretary of State for Education. All subjects have to receive acceptance from this committee before they can be taught, as without acceptance pupils would be unable to gain a certificated qualification. Many Regional Examination groups have adopted a design for the front cover of subject booklets which includes a 'rubber stamp' design, 'SEC Approved'. This was done as much to attract schools to adopt their subjects for examination in competition with the other Regional Examination Groups as to inform a reader that the document was official.

GENERAL CRITERIA

The General Criteria allow for the opportunity of Mode 2 and Mode 3 syllabuses to be submitted for approval. This is a development that was very much used by the CSE Boards and has been considered as a necessary innovation to be continued.

As with subjects that have been accepted or approved by the SEC under the heading Subject-specific Criteria, the presentation of aims, objectives, syllabus, etc. must be in a format that is very similar. It is certain, however, that the Regional Examination Groups will do all they can to avoid the proliferation of subject titles and syllabuses. Considerable time and effort have been given to the development of subjects under the heading of Subject-specific Criteria; it is anticipated that most needs are catered for and that there will not be a need for alternative courses to be submitted.

Having said this, it is important that an avenue still exists for subjects to be submitted which contain elements non-existent in those that have already been approved, and which will be appropriate and beneficial to even a small minority. The aim will naturally be to cater for everyone but not to create a situation which swamped the CSE Boards and caused considerable confusion for employers, who could not interpret the results in terms of suitability of applicants for a particular job or career.

SUBJECT-SPECIFIC CRITERIA

At the time of writing, twenty subject areas have been approved, one of which is Craft, Design and Technology. All aspects that have been referred to in the past as having a technical bias will carry this title. At present three aspects have been approved. They are:

CDT – Design and Realisation
CDT – Technology
CDT – Design and Communication

Because one of the important aims of the GCSE has been to cut down the proliferation of titles, these three replace such titles as: Woodwork, Metalwork, Handicraft, Design Studies, Design Communication and Implementation, Design Craft Wood, Design Craft Metal, Technical Studies, Technical Drawing, Engineering and Geometrical Drawing, Graphic Communication, Project Technology, etc. From just this list, which is by no means complete, it is easy to see how the proliferation of titles can, and has in the past, caused confusion. The reduction in the number of titles should also be less confusing for pupils making their course choices in the third year.

The refining and sifting operation carried out to reduce the number of subject titles is only the beginning of a major change. For if we look closely at the aims and objectives we will see that they are common to each of the three strands – Design and Realisation, Technology, and Design and Communication – thus doing away with the need for a slightly different set of aims and objectives for each of the strands and hence a proliferation of reading material.

AIMS

As with the syllabuses published for subjects in GCE and CSE, a statement of aims heads the details to be given in any syllabus. These have already been established for CDT and read as shown in Figure 1.

Assessment objectives

Each assessment objective (see Figure 2) is given a number which serves essentially as a reference number. The advantage of this will be appreciated later, in the section on the relationship between Assessment Objectives and content. Other uses will also become apparent when teachers are faced with the task of planning lessons, when the Objectives will serve as a valuable checklist. The numbers, however, do not indicate a sequence or order of importance. All objectives are important, and it is vital that they are not missed.

The Assessment Objectives are written in such a way that it is possible to measure different levels of performance, and this will be seen in the section on grade-related criteria.

It is important to realise that there is a limited degree of flexibility in the wording of these Assessment Objectives and

▲   *Fig. 1 Aims of the Courses*

▲   *Fig. 2 Assessment Objectives*

that there may be some slight difference from one Regional
Examination Group to another and from one strand to another.
So it is important to read these carefully. If we read Assessment
Objective number 14 and compare the wording of the same
number Assessment Objective published by the Southern
Examination Group for their CDT Design and Communica-
tion, we will see that some modification has taken place without
really changing the meaning of the original version.

> **14** propose or make modifications to a product or system both
> during manufacture or modelling, and after completion and
> evaluation; or where more appropriate, at the drawing board
> stage of manufacture. (SEG)

It would only be fair to point out that this was the only
amendment and that all the other Assessment Objectives were
identical with those in Figure 1.

What goes on in the classroom, workshop or design studio is
going to be very much influenced by these objectives, and for
some teachers quite a new approach will have to be adopted if

these objectives are to be achieved. The changes could well start with the reorganisation of equipment, tools and materials. The Woodwork room, with its ten woodwork benches, two wood turning lathes and drilling machines, is not going to be adequate or appropriate for a pupil wanting to produce an artefact in more than one material or 'demonstrate graphical and other communication skills necessary to give, in a clear and appropriate form, information about an artefact or system' (Assessment Objective 2). The drawing office will also have to take on a new look if candidates pursuing CDT design and communication are going to 'demonstrate appropriate skills, make or model the artefact or system' (Assessment Objective 13).

These changes are only the tip of the iceberg. It is inevitable that many teachers will have to forgo their love of craftmanship for craftmanship's sake and consider their responsibilities in the light of being teachers of Design and all its implications. The ability to make is still important and is covered in Assessment Objective 13, but balanced with the other associated activities it plays a far less important role than the part it played in those craft-based subjects – Woodwork, Metalwork offered at GCE and CSE level.

Fortunately many of the GCE Boards and CSE Boards introduced 'design-based' subjects more than fifteen years ago, so providing the opportunity for teachers to educate candidates through 'problem-solving' activities. For them the introduction of GCSE CDT will be a natural progression from the work that they have been doing, and an assurance that what they have been doing was of sufficient importance for the present examination to be modelled on the example that they have set. The statement made in the SEC's Working Paper No.2, 'The aim in each subject should be one of making that which is important measurable rather than making what is measurable important', can now be fully realised in CDT courses by following the seventeen Assessment Objectives.

CONTENT

As with every subject there is a need to state the minimum amount of knowledge to be learnt and the activities to be performed. The essential activity of all CDT evolves round designing and problem-solving which leads to a solution and evaluation. This is now a well-established ingredient of many GCE and CSE design-based courses and should be familiar to most subject teachers. However, those who have been mainly or solely involved with such subjects as Technical Drawing, Geometrical and Engineering Drawing, etc. will need to examine the Assessment Objectives very carefully, for now the area of communication which plays a large part in the activity of designing artefacts or solving problems is seen as an essential ingredient of any drawing course. As with the practical subjects of the past, where now it is no longer sufficient to just manufacture artefacts, so it is no longer sufficient for pupils to be engaged in contrived activities such as constructing an ellipse or the development of a square-based pyramid, unless the activity can be related to the production of a model that is a solution to a problem. In such a case learning how to construct the ellipse or the square-based pyramid not only assisted in the arrival of a solution but also gave a justifiable reason for acquiring such specialised skill and knowledge.

Though candidates are expected to acquire a level of knowledge, it is envisaged that this will be mainly gained as a result of investigation during designing and solving problems and the skills of making during the construction of the solutions. In this way the purpose of learning, it is hoped, will be seen as a means to an end rather than just a need to learn unrelated pieces of knowledge in order to pass an examination. Learning about safety should also be relevant to the workshop situations experienced by the pupils and not merely a copying exercise of rules and regulations, sometimes given as a punishment. This is not good practice and has certainly not proved to be a successful way of learning.

For definitive information about the area of knowledge, the teacher is advised to read the documents published by the Regional Examination Groups. Broadly speaking it is more important to be familiar with marketing forms of material than it is to know how timber is seasoned or iron and steel are produced. Children are no longer expected to reproduce cross-section drawings of timber being kiln-seasoned or furnaces of all kinds that are used in the production of metals. Instead they are expected to apply knowledge to given situations. With the information and knowledge of the various marketing forms of materials, whether it be wood, metal or plastics, the pupil is partially equipped with the information that is needed when selecting a material for the realisation of a problem. It is important to emphasise at this stage that the written papers are designed to test the pupil's ability to apply knowledge rather than the ability to regurgitate information. A closer examination of the types of questions now posed in examination papers (pages 32–6) will help the teacher to recognise these points.

RELATIONSHIP BETWEEN ASSESSMENT OBJECTIVES AND CONTENT

Assessment Objectives have been clearly stated in the *National Criteria – Craft Design and Technology*, issued by Department of Education and Science, Welsh Office, and can be found in Figure 1.

The days of entering, as a visitor, a workshop/design studio full of children involved in a design activity and being able to identify clearly what educational objective is being achieved by the class have gone. The children could be involved in any one of three main areas:

1 The acquisition of design skills;
2 The acquisition of subject-related skills – principally making and assembling in Design and Realisation, and Technology, and principally communicating and making in communication;
3 the acquisition and application of knowledge.

Then within these three main areas the pupils could be involved in any one of the 17 educational objectives given in Figure 1. This may at first appear to be impossible to organise and control, but in a well-managed situation the pupils become the selectors and organisers of their activities, while the teacher adopts the role of the consultant.

Some Regional Examination Groups present the syllabus content alongside the Educational Objectives (see London and East Anglian Group) which makes it possible for the teacher to relate an activity with an educational objective very easily. The guidance notes which occupy a third column provide further assistance to the teacher and help to give further reassurance.

Again it will be necessary for the teacher to be knowledgeable of the mark weighting or value given to each of the three areas by consulting the documents the Groups publish. For although an analysis is given in the next chapter, there is nothing to prevent the Regional Examination Groups from adjusting the weighting within the boundaries given if they so wish. The mark weighting expressed as a percentage must fall within the following boundaries:

design skills	40 – 45%
subject-related skills	30 – 40%
knowledge	20 – 25%

but, whatever weighting is given, the total of the three must add up to 100 per cent. The Assessment Objectives contained within these areas are unlikely to change, and at present are as follows:

	Assessment Objectives
1 design skills	1, 3–7, 9–11, 14–16
2 subject-related skills	1, 2, 4, 6–8, 11–16
3 knowledge	1, 4–6, 10, 11, 14–17

As can readily be seen, some Assessment Objectives appear more than once and many of them appear in each of the three areas, e.g. Assessment Objective 1, 'describe and apply facts, principles and concepts related to artefact and/or systems design, realisation and evaluation'. The objective is applicable to all three areas, and like other objectives which appear more than once, helps to underline the fact that the boundaries of each are not confined and fixed to a given area. Hence at the end of a course many of the objectives will be assessed in the course-work, design project and written paper. This will not mean a duplication of work, but rather a reinforcement of the importance of the objectives being assessed.

TECHNIQUES OF ASSESSMENT

To a large extent, the decision as to how candidates will be assessed will be within the control of the Regional Examination Groups. They will be responsible for the appointment of Chief Examiners, Moderators and the necessary assistants to conduct the process of assessment and ultimately the process of awarding grades. Therefore in the areas of marking written papers, design assignments set by the Regional Examining Group will all be conducted by those appointed to do so.

However, in the area of coursework the pupil's own teacher will be involved, very much in the same way as has been the practice in some GCE O level and CSE subjects that contained a practical element. The teacher's role is vital in the assessment process and cannot be replaced by an externally appointed person, for it is only the candidate's teacher who can assess the contribution made by the candidate in the designing and making of an artefact. Some aspects of the resultant work may have been largely due to the teacher's effort to sustain the interest of a poorly motivated pupil. In which case, for a similar resultant parcel of work in which the candidate made a major contribution, credit should be awarded. It is anticipated that some projects may result from a team effort and it will be the teacher's task to know and assess the contribution made by each individual. As in these examples given and other similar situations, it is clear that the first stage of the assessment procedure must be conducted by the candidate's own teacher.

Differentiation

'A differentiated approach to assessment must be adopted in all syllabuses to ensure that all candidates have the opportunity to demonstrate what they know and can do.' With this criterion in mind it is not difficult to appreciate the importance of coursework, where the candidate is largely responsible for selecting the work that he or she can do. The least able are much more likely to tackle a problem to which there are known solutions and which only requires basic practical skills for production, while at the other end of the ability range candidates are more likely to choose a project to which there are no obvious solutions and revel in the challenge to produce something exciting and interesting. This natural selection of differentiation through coursework is a very reliable means of identifying what can be done at all levels. Here again the teacher has an important role to play in guiding the pupils. Keeping the less-able pupil away from an overambitious project, while at the same time encouraging a less confident but more able pupil to tackle the more demanding and open-ended problems, is of vital importance and crucial to determining grades of attainment.

There are other ways in which differentiation may manifest itself all along the route from the identification of a problem to the evaluation of the solution. The less-able pupil may or may not identify a problem, and will require a problem to be presented—often by the teacher. The more able pupil will have recognised a need for a particular situation and been able to put into words a 'design brief'. Even if, by coincidence, two pupils are tackling an identical problem, the one who has produced the design brief unaided has obviously demonstrated a difference of ability from the one who needed assistance. While, for the purpose of demonstrating that coursework is a valuable aid to allow differentiation to manifest itself by choosing examples at the extremities of the examinable ability range, it is not too difficult to identify levels of variation so that assessments can be translated into marks. This aspect will be looked at more closely in the next section under the heading Grade Descriptions.

The written papers are also presented in such a way that candidates are able to show what they know rather than what they do not know. A paper can be designed with a particular level of ability in mind. These are usually at the extremities of the ability range, and some subjects have opted for this mode of differentiation. Others have opted for a common paper containing differentiated questions in which there is a choice of question, and some questions are open-ended. A review of sample papers in the next chapter should assist the reader in identifying the modes of differentiation that are being adopted by the Regional Examination Groups for CDT.

Project work

A compulsory project which requires all candidates to attempt the same problem or to attempt one problem from a restricted choice of approximately three is yet one more technique used to bring about a difference in the levels of work. Since the projects are set by the examining body, this allows for a control to be used when work has to be moderated. This has been a technique employed by GCE and CSE Boards for a number of years. In the early days of setting design problems the candidate was required to answer a problem under the formal conditions of a traditional type of examination. In the space of approximately two hours a candidate was expected to read a question paper, select a problem and then follow a sequence of operations, i.e. analysing the problem, developing ideas by sketching, and producing a working drawing, without being able to refer to reference material that is so essential when 'designing'. Though certain skills could be assessed, these were limited by the restrictions of a formal examination, and in the light of experience a much improved system has been developed.

All candidates receive a problem, or set of problems, 12–15 weeks before the date of completion. This enables them to think around the problems before making a selection, and once a problem has been chosen, to begin researching for material that will assist them with resolving the problem. They can do this very thoroughly and document their findings in a design booklet which is often provided by the Examination Group. The conscientious candidate will have every opportunity to show the depth of enquiry that has been carried out and be differentiated from the less-well motivated candidate very easily. The more able candidate will also have the opportunity to demonstrate communicative skills by using a variety of illustrative techniques, from quick sketches in pencil to a detailed two-point perspective view rendered in pen and ink. Because of the removal of the restrictions of a formal examination, time can also be spent on using airbrush techniques, should the candidate wish to do so. The Design booklet, which represents the culmination of all the research and design thinking, becomes one of the instruments by which the candidate is assessed, and replaces the formal Design examination that has been used by many GCE and CSE Examination Boards.

Practical examination

This progressive move leads on very nicely to the next instrument of assessment which is the practical examination. For years candidates have been exposed to the tortuous experience of making a useless artefact with all the restrictions that are imposed by a formal examination in conditions that often generate a state of panic when a piece of wood splits or the wrong piece has been removed. These conditions are not only alien to the usual experience a pupil has in a workshop, but they are also alien to showing what a candidate can really do. Thankfully, very few of these situations exist today and more enlightened approaches have been adopted. For those teachers who have been more fortunate in being able to enter candidates in the CDT subjects with the more enlightened approaches in the GCE and CSE examinations there should be no problem in moving towards the approaches adopted in the GCSE CDT subjects.

The change that is most likely to be observed first is the absence of a practical examination paper giving details of what has to be made under examination conditions. In its place will be a practical examination where the candidate will be required to make the artefact that has been designed by the candidate during the 12–15 week period of preparation. The details on the working drawing will be the result of the candidate's own investigation and development of ideas, selection of ideas and decisions about material, construction, assembly and finish. The teacher will, of course, have the opportunity of amending the candidate's ideas, if necessary, to ensure that making the design is a viable proposition. Such amendments will have to be relayed to the examining body so that a complete and accurate account of what has taken place can be considered at the time of assessment.

To summarise – differentiation will be achieved through the candidate performing in coursework carried out during the first four terms, the compulsory project in term five, the practical examination and the written examination taken in term six of the two-year course. This is likely to be the format adopted by most

Regional Examination Groups, but a slight variation must be expected.

GRADE DESCRIPTION

Some subjects have already been using grade description techniques to determine what has been achieved by a candidate and found the technique to be very fair, objective and reliable. The well-established technique of norm referencing is not primarily intended to show individual levels of attainment, and has thus proved to be unreliable and not very helpful when it comes to wanting to know what has been achieved. With the GCSE's underlying philosophy of providing an opportunity for pupils to show what they 'know and can do', it is only appropriate and reasonable to expect that a system should be adopted to identify what has been achieved at all levels, from A to G.

The expectations of achievement will not only have to be related to the objectives outlined earlier in Figure 2, but also to a grade that will indicate a level of competence. This has been attempted by working parties and will continue for many years. Despite the experience, expertise and knowledge that these working parties embody, it is almost impossible to arrive at a description that fits perfectly the performance expected of a grade C candidate, or any other candidate for that matter. However, the need to arrive at a set of Grade Descriptions that relate to the assessment objectives is crucial as a measuring instrument to be used in the final stage of the assessment process.

For the most part it will be the Examiners and Moderators who will have to apply grade descriptions to the candidates' work, but a major and vital role will have to be played by the candidate's own teacher, who will be responsible for the initial assessment of the coursework element. Details of the methods to be employed by teachers will be given in Chapter 5.

2 The Syllabuses

REORGANISATION OF THE SECONDARY SCHOOL EXAMINATION SYSTEM

Though the reader may be familiar with the setting up of the six Regional Examination Groups a list here should help identification of the details given in the tables that follow. For further information, reference can be made at the back of the book where details are given of the GCE and CSE Examining Boards' grouping and addresses.

London and East Anglian Group	LEAG
Midland Examining Group	MEG
Northern Examination Association	NEA
Northern Ireland Secondary Examination Council	NISEC
Southern Examination Group	SEG
Welsh Joint Education Committee	WJEC

Mark Allocation

The three main areas of learning that were mentioned in Chapter 1 were: designing – making and assembling in Design and Realisation, and also in Technology; communication, in Design and Communication; and acquiring knowledge. For each of the three areas the examining bodies calculated independently of each other the weighting that should be allocated to each area, working of course within the boundaries stipulated by the Department of Education and Science and the Welsh Office, the details of which are presented in Chapter 1.

▼ Fig. 1 Tables showing mark allocation made by examining groups for the three strands of CDT

DESIGN & REALISATION

GROUP	DESIGN SKILLS 40–45	SUBJECT RELATED 30–40	KNOWLEDGE 20–25
LEAG	40	35	25
MEG	40	40	20
NEA	45	35	20
NISEC	40	40	20
SEG	45	30	25
WJEC	40	30	25

TECHNOLOGY

GROUP	DESIGN SKILLS 40–45	SUBJECT RELATED 30–40	KNOWLEDGE 20–25
LEAG	40	35	25
MEG	45	35	20
NEA	40	40	20
NISEC	40	35	25
SEG	40	35	25
WJEC	40	35	25

	GROUP	DESIGN SKILLS 40–45	SUBJECT RELATED 30–40	KNOWLEDGE 20–25
DESIGN & COMMUNICATION	LEAG	40	35	25
	MEG	40	35	25
	NEA	40	40	20
	NISEC	*		
	SEG	40	40	20
	WJEC	40	35	25

*SEC Approved copy not available at time of writing

▲ *Fig. 1 Continued*

These tables provide an indication of the emphases each Examination Group adopted and a quick comparison can be made. However, it is necessary to have a further breakdown of mark allocation to understand a little more fully. This can be done by looking at the assessment patterns. They are usually presented in tabular form and look similar to the example illustrated in Figure 2.

	DESIGN	SUBJECT Related skills	KNOWLEDGE	TOTAL
Assessment Objectives	1, 4, 5 6, 8, 10, 14	7, 9, 10 11, 12	2, 3, 13	
Paper 1 Written	5%	10%	15%	30%
Paper 2 Design Assignment	20%	5%	5%	30%
Coursework (including mini project)	15%	20%	5%	40%
TOTAL	40%	35%	25%	100%

▲ *Fig. 2 Assessment pattern for LEAG Design and Communication*

Assessment Patterns

The assessment patterns presented by each of the six Regional Examination Groups have been set out in a combined table to aid cross-referencing, see Figures 3, 4 and 5. By just glancing at these tables it is possible to glean a little more information and to begin to recognise where likenesses and differences occur. For example, it is possible to single out those courses that are biased towards the work that is to be assessed as coursework; (the maximum allowable being 50 per cent). Just as it is possible to note those courses that favour assessment by written examination.

This quick and ready way of cross-referencing on a table is helpful when faced with the task of selecting either the Regional Examining Group or the course that will be most suitable for the pupils' needs. However, before a final decision is made it would be most prudent to consult a detailed syllabus and assessment

Assessment of weighting

GROUP	COURSE WORK	WRITTEN	DESIGN	DESIGN & REALISATION	
LEAG	40%	30%	–	15%	15%
MEG	30%	30%	–	20%	20%
NEA	50%	20%	30%	–	–
NISEC	50%	25%	25%	–	–
SEG	50%	25%	25%	–	–
WJEC	50%	30%	20%	–	–

Time allocations

COURSE WORK	WRITTEN	DESIGN	DESIGN &	REALISATION
2 yrs	2hrs	–	20th Jan-20th Mar	Max 6 hrs
2 yrs	2¼ hrs	–	Dec-1st Mar	Max 20 hrs
1yr prior to exam	1¾ hrs	Research Mar-Exam 1¾ hrs	–	–
2 yrs	2 hrs	2 hrs	–	–
2 yrs	2 hrs	Research 6-8 hrs Exam 2 hrs	–	–
11 mnths	2 hrs	3 hrs	–	–

DESIGN & COMMUNICATION

▲ *Fig. 3 Assessment patterns for Design and Realisation*   ▼ *Fig. 4 Assessment patterns for Technology*

TECHNOLOGY

GROUP	PROJECT	CORE WRITTEN 1	MODULE WRITTEN 2	DESIGN OR EXTENSION
LEAG	50%	30%		20%
MEG	A 60%	20%	20%	
	B 50%	30%		20%
NEA	50%	25%	25%	–
NISEC	45%	20%	15%	Ext'n 20%
SEG	50%	20%	30%	–
WJEC	50%	20%	30%	–

PROJECT	CORE WRITTEN	MODULE WRITTEN	DESIGN EXTENSION
45 hrs	2¼ hrs		2 hrs
A	1½ hrs	1½ hrs	
B	2½ hrs		2½ hrs
45 hrs	1½ hrs	1½ hrs	–
Final Year	1½ hrs	1 hr	Ext'n 1 hr
Final Year	1½ hrs	2 hrs	–
2 years	1½ hrs	3 hrs	–

* A recommended for candidates expected to achieve a grade equivalent to GCE grade C and below.
  B recommended for candidates expected to achieve a grade equivalent to GCE grade C and above.

DESIGN & REALISATION

GROUP	COURSE WORK	PAPER 1	PAPER 2	PAPER 3
LEAG	40%	30%	30%	
MEG	40%	25%	35%	
NEA	33%	37%	30%	
†NISEC	20%	40%	40%	40%
SEG	50%	20%	30%	
WJEC	30%	30%	40%	

COURSE WORK	PAPER 1	PAPER 2	PAPER 3
½ term 3½ terms	2½ hrs	Term 5	
Terms 2, 3, 4 & 5.	2¼ hrs	2¾hrs	
45-60 hrs	2½ hrs	2½ hrs	
During the course	1½ hrs	1½ hrs	2½ hrs
Final year	1¾ hrs	2 hrs	
11 months	2 hrs	2½ hrs	

†TECHNICAL DESIGN AND GRAPHIC COMMUNICATION
Candidates to be considered for an award in the C-G range should take Papers 1 and 2 plus Coursework.
Candidates to be considered for an A-E range award should take Papers 2 and 3 plus Coursework.

▲ *Fig. 5 Assessment patterns for Design and Communication*

scheme. The real differences can only be picked out by a close scrutiny of all that is published by the examination group. Besides the published syllabuses, Teachers Guides are also printed to help the teacher understand more fully what the course is all about and, most important, how to set about assessment procedures.

Selecting Courses

Many teachers are starting a basic CDT course for their pupils without committing themselves or their pupils to one particular strand. For they have recognised that the central theme of solving problems is common to each strand and that an introductory course can be equally suitable for each of the three strands. Leaving the option open until such time that entries have to be submitted does allow time for pupils to demonstrate their interests and potential development before they are committed to being assessed in a strand for which they might be unsuited. Some courses require all the work that has been covered during the two years to be submitted for assessment while others, like the Design and Realisation offered by NEA, only require that work which was carried out in the twelve-month period leading up to the examination be submitted, thus leaving those teachers who choose to enter their pupils for such a course an opportunity to keep their options open for at least two full terms.

Where courses are shown to require two years' work for assessment on the table, it is worth examining the finer details given in the information published by the examining body; for in the case of the Design and Realisation course offered by LEAG it would appear that all the work done during the two years is required for assessment when in fact only a sample selected by the pupil is required. In the normal uninterrupted course of events the work carried out in the final year is often the pupil's best work and is therefore most likely to be selected for assessment.

The problem of selecting between strands is not likely to cause much heart-searching since the pupils generally know which area of CDT they most enjoy. They may make their judgements on the basis of whether they can draw well or whether they are interested in electronics and mechanics and so on, or whether they like to be creative. These may not be well-informed judgements, but may be reliable indicators as to which course is suitable.

The final decision must be made by the teacher, but a decision arrived at by a mutual agreement between pupil and teacher has the greatest chance of being successful.

Course Emphasis

All three CDT strands provide independent courses, but most of their objectives have much in common. The relationship can be shown diagramatically.

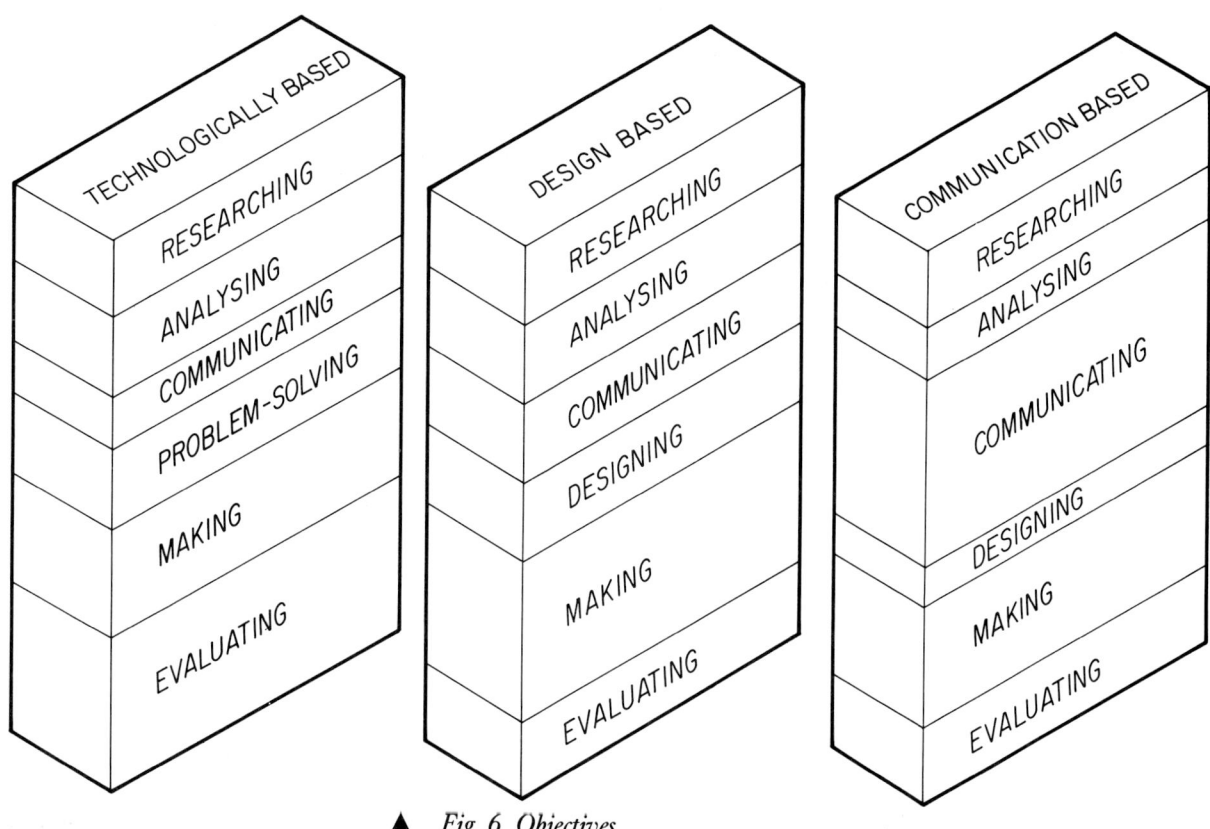

▲   *Fig. 6 Objectives*

All six Regional Examining Groups reflect the same pattern within their syllabuses; the emphasis placed upon each objective for each of the three strands can be clearly seen in the diagram above. There is a greater emphasis placed upon the communication element for 'Design and Communication' than for either of the other strands.

	Design and Realisation	Technology	Design and communication
Skills (virtually identical statements)	design making communication	design making communication	design making communication
Knowledge	materials and components	materials and components	materials and components
	principles and concepts		principles and concepts
		energy	energy
		control	control
	design and society	technology and society	graphic design and society

▲   *Fig. 7 Course component matrix from SEC's 'A Guide for Teachers'*

Similarly, as might be expected, the element of making plays a lesser role in both the Technology strand and the Design and Communication strand than for Design and Realisation. The bias in emphasis should also emerge in the course content of each of the three strands but with such considerable common ground the differences occur mainly within the context of its use. Figure 7 illustrates clearly the common ground and the bias of emphasis in each of the three strands.

The Common Core

Below is an analysis of the knowledge content 'Common Core' taken from the Design and Realisation syllabuses offered by the six Examining Groups.

Areas of knowledge
1 Material
    Forms
    Characteristics
    Properties
    Appropriate uses
    Comparative qualities
    Economic factors
2 Working materials
    Construction and application
    Specific processes, ie finishing
    Safe practices
    Fixings and assembly
    Surface treatment
    Tools and selection for use
    Processes and techniques
3 Principles and concepts
    Energy
    Control
    Movement mechanisms and motion
    Adjustment
    Holding
    Locking
    Joining
    Forces and structures
    Wear
    Ergonomics and anthropometrics
    Aesthetics
    Design and society

The depth to which any of the topics above are taken and the amount of detail covered vary considerably, but there appears to be a balance in the end.

Compare the following extracts taken from two Design and Realisation syllabuses:

Syllabus A

Themes

Candidates should develop a knowledge base in each of the following themes taken from the National Criteria for CDT. They are interwoven into all the aspects of Design and Realisation activity. (Whilst pupils may not gain sufficient experience to recognise and understand the principles and concepts behind the themes, they should be able to give examples of each drawn from their own design and making experience.)

The notes below suggest ways in which the themes could be covered during a course in Design and Realisation.

Energy Candidates should appreciate simple energy transformations and be aware of the main forms in which energy is used and conserved. The ways in which Design and Realisation activity highlight the use of energy should lead to some appreciation of mankind's dependence on energy (past, present and future) to achieve personal and social purposes.

Control This concept impinges on much Design and Realisation activity, principally in simple forms such as the static control achieved by fastenings, structures and holding devices. Repeat work, considerations of multiple production, any use of precision machinery, any jig or fixture, and many other common Design and Realisation situations should help all candidates to recognise examples of this fundamental and wide-reaching concept. (Candidates engaged in work with a technological bias will meet more involved examples.)

Movement Pupils should understand simple relationships between moving parts of machines and mechanisms. Toys, locks, bicycles, vehicles, mobiles, hinges, display systems, lighting effects and other arrangements illustrate situations by which simple principles of movement can be understood.

Adjustment Through various experiences concerning tools, materials, processes and design considerations, pupils should have an overall appreciation of the necessity, purpose and incidence of adjustment as a concept widespread in Design and Realisation activity.

Holding The wide spectrum of devices and arrangements from the human hand to the three jaw chuck, pins to sash cramps, and purses to bookshelves which is encountered constantly in Design and Realisation activity should enable this concept to be widely understood.

Locking As well as overlap with the concept of control, candidates should recognise locking and inter-locking in Design and Realisation activity with connotations of fit, precision, connections, matching.

Joining Any use of materials will provide numerous instances from which a conceptual understanding of joining can be built. The more varied the experience and the more widespread the materials, the better a candidate will be able to appreciate the principles involved such as surface area of contact, compatability of material, properties of fastenings and fixings, basic conditions of adhesion.

Forces and structures An elementary understanding of the common forces found in structures and other situations should be assimilated by pupils in a Design and Realisation course.

Compression – as in a table or chair leg.
Tension – as in a frame saw blade or hot wire cutter.
Torsion – as in a key when turned in a lock.
Leverage – as in the use of spanner or pincers.
Shear – as in the use of a guillotine or scissors.

Wear This concept can be understood at various levels. It should include the physical characteristics of materials (especially hardness), processes of friction and lubrication, operations of abrasions and wasting.

Surface treatment Candidates should consider the technical and aesthetic aspects of surface treatment under this heading. The economic and social significance of surface treatment in consumer products and environments should be appreciated as well as the technical and aesthetic considerations that arise from the candidates' own work.

Ergonomics Candidates will be expected to understand the importance of ergonomic considerations in any design situation. The level of application will necessarily be simple and direct in many projects, but the principle of taking such considerations into account is fundamental for successful designing.

Aesthetics The assessment of the project will take appropriate aesthetic qualities into account. Candidates should, as a minimum, understand what is meant by terms such as form, shape, line, proportion and balance and be aware of colour and the basic aesthetic potential of the materials in which they work. The effects of workmanship, material combinations, texture and degree of finish upon aesthetic quality should be appreciated. Candidates should enlarge their appreciation and understanding of the aesthetic dimension by considering the qualities of industrially made products.

Design and Society Candidates should be aware of a relationship between their own work and its counterpart in the industrial and social life of the world around them. This awareness may be enhanced through issues such as mass production, consumer awareness, conservation in its many forms, pollution and some possible futures.

New Technology
The use of such technology as the microprocessor and CAD/CAM is to be encouraged in Design and Realisation courses although questions in this area will not be set. Candidates who do, however, refer to or apply technological developments will receive credit where the reference or application is appropriate.

Syllabus B

Principles and Concepts
Simple concepts relating to energy, control movement, holding,

locking, joining, forces and structures. Wear, surface treatment. Ergonomics, aesthetics.

Now compare two more extracts taken from the same two syllabuses.

Syllabus A

Knowledge

Materials
Candidates should be familiar with the following aspects of the main materials they have studied.
(a) availability and form of supply; costing and comparability with other materials
(b) physical characteristics
(c) suitability for particular purposes
(d) methods of working or forming the material
(e) methods of joining the material to similar and to dissimilar materials
(f) appropriate surface treatments and finishes
(g) codes of safe practice
(h) ancillary materials associated with main materials as appropriate (e.g. solders, adhesives, lubricants, abrasives)
(i) fixing devices and fastenings associated with main materials (e.g. screws, hinges, pop rivets, velcro)

The processes and skills associated with the correct choice and use of tools and equipment will be tested only through the realisation of coursework.

Syllabus B

The following should be taught in the context of design and making.

Knowledge

Materials and Components	**Candidates should have a fundamental knowledge and understanding of wood, metal and plastics, to include the following:**
Characteristics	— the general classification of materials i.e. ferrous/non-ferrous metals, hard/soft-wood, manufactured boards, thermo/thermosetting plastics.
Working properties	— the making of simple comparisons between these materials in relation to strength, hardness, toughness, weight, plasticity, durability, aesthetic qualities.
Market forms	— the shapes, sizes (general not particular) and sections in which these materials are available and comparative knowlede of their cost. Common/specialist applications ancillary materials (e.g. adhesives, solders, lubricants, abrasives, finishes, fixing devices, fastenings).
Process	— the correct tools for preparation, marking out measurement and checking.

| **Tools and equipment** | — the hand and machine tools and equipment which are required for these processes and their correct use. |

It can be clearly seen that a variation in the depth of information can be balanced by more detail given in another section. However, to gain a fully comprehensive understanding of what is required by pupils and teachers following a particular course it would seem necessary to read the details given by more than a single Regional Examining Group. Being armed with detailed information from a number of syllabuses certain gaps will be filled and the teacher is better equipped to develop schemes of work.

ASSESSMENT FORMAT

In terms of guidance for the CDT examination format, the *National Criteria* were not too directed or prescriptive. In analysis of the routes chosen by the Examining Groups there is a high degree of correlation, as the following summary shows.

The advice from SEC suggested the following modes as desirable:

a) A common examination with the proviso that differentiated papers might be applicable for testing the attainment of candidates at the extremities of the ability range (para. 8.1). *All Examining Groups have opted for a common core examination in all CDT strands.*

b) A combination of formal written papers and assessment of coursework, so that design skills, subject-related skills and knowledge are tested (para. 8.2). *All Examining Groups have accepted this format in all CDT strands.*

c) Differentiation and related target grouping can be achieved with written papers by a choice of questions, differentiated questions and open ended questions (para. 8.1). *This principle has been accepted by all Groups.*

d) Examination structure (para. 8.3): Examination papers having an incline of difficulty. Also the possibility of having an incline of difficulty within a question. *Whilst several of the Examining Groups have used this structured technique, others have used 'targeting by grids' techniques. There was no evidence of a random approach. All showed structure in their format.*

f) The use of pre-printed/answer booklets is desirable. *Whilst the layout of some papers had not been considered in totality, all Examining Groups are using combination question/answer booklets.*

g) Adequate time to research design briefs which are later required to be realised or modelled. *In all cases where candidates were required to proceed beyond the design stage, time for research was included.*

h) and j) 'Design then make situations' are suitable and

appropriate for external assessment. *Examining Groups applied this approach in the form of projects, or alternatively as an externally assessed element of coursework. Further, several used a 'thematic' approach which is made known to candidates well in advance of the examination date so that research time is available.*

k) The need for teacher involvement in course/project work assessment. *All examining groups placed considerable emphasis on and gave guidance for this aspect of assessment.*

m) Candidate folios for research and development of projects to include mock-ups and models. *There was positive encouragement and recognition to the candidate for these aspects of the examination throughout all Examining Groups across all strands.*

When analysing the take-up of those aspects considered undesirable by the *National Criteria*, there was no evidence that any examining group intended to involve:

8.3 c) Tariff questions.
d) Open-ended essay questions for factual knowledge, i.e. completely blank pages in the answer book.
l) A timed externally set test piece.

Taking into consideration the constraints and complexities involved, it is not surprising that little evidence was apparent for:

8.3 d) Oral questions and responses though costly in terms of time and other resources might have a place in some assessment procedures.

Such assessment will be applied by most schools in their coursework assessment plans, particularly when continuous assessment is required. In terms of the problems schools face in respect of coursework required for moderation, most groups have been surprisingly realistic and are prepared to accept 'authenticated' photographic evidence in lieu of the artefact. Where the facilities are available, many schools are preparing to use edited video recordings in support of a candidate's folio work. This is applicable particularly where evaluation and refinement form part of the designing activity.

Marking Written Scripts

An analysis of the approaches adopted by the six Examining Groups shows that two contrasting styles emerge. In the first approach three principles are applied:

1 Several marks are allocated for each component of an answer.
2 A positive build up of marks is made for every acceptable response within the mark allocation.
3 It is assumed that there is more than one acceptable answer to a question.

For example, each question is allocated a minimum of 2 marks no matter how short a response is required. This allows for even a vague response to gain credit if it is deserved, and for the more specific answer to get even more credit. With a question so designed to elicit a number of pieces of information from a candidate it is possible for an accumulation of marks to be achieved for every acceptable response made. Finally, the questions may be couched in such terms that there is more than one acceptable answer. To answer the question 'Describe a method and name the tools used to draw a parallel line to a long straight edge of a piece of resistent material' the response will be very much influenced by the material chosen and quite different techniques will be employed. The figure below illustrates this.

▼ *Fig. 8 Specimen marking scheme, LEAG*

Specimen Marking Scheme

Question	Requirement	Response	Mark Scheme Application
A 1 *(a)*	Tool name(s)	Pen/ruler Pencil/thumb gauge Wood gauge	1 mark (Does the job) 3 marks (Job, name, completeness) 2 marks
		and so on according to response	
A 3 *(c)*	Adhesive laminate to wood	Evostik Evostik Resin W Contact adhesive Laminate glue	1 mark (Group, name, bonding action unclear) 2 marks – unless cramping mentioned 2 marks (No proprietary name given) 1 mark
		and so on according to response	
A 7 *(a)*	Safety turning between centres (wood lathe)	Chuck key flying out (remove key before use) Drill breaking (take care) Hair caught in chuck (use cap, and chuck guard)	1 mark Chuck guard should obviate risk 1 mark for risk operation 4 marks
		and so on according to response	

A 1. *Selection of marking tools*

Criteria	Does the tool do the job?	1	
	Does it do the job properly?	1	
	Is it correctly named?	1	
	Is it completely named?	1 $= 4 \times 2$	8

A 2. *Property of tool steel*

Criteria	Is property	*(a)*	True for tool steel	1	
		(b)	Applicable to edge tools	1	
		(c)	Correctly termed?	1	3

A 3. *(a) Adhesive for acrylic   (b) softwood   (c) laminate to wood*

Criteria	Proprietary name	1	
	Technical group name	1	
	Bonding action	1 $= 3 \times 3$	9

A 4. *(a), (b), (c) surface finish,*

Criteria	Is tool for cutting to shape/size mentioned	1		
	Is equipment used to obtain smooth surface mentioned?	2		
	Are various grades mentioned?	1	$= 4 \times 3$	12

A 5. *Speed of drilling machine*

Criteria	Do the pulleys match?	2	
	Have the correct pair been chosen?	4	6

A 6. *(a), (b), (c) Cutting tools for material/process*

Criteria	Is the tool suitable for the material?	1		
	Is the tool suitable for the process?	1		
	Has the correct tool category been given?	1		
	Has a full specification been given?	1	$= 4 \times 9$	36

In contrast, the marking scheme illustrated on page 32 is typical of the other pattern. Here there has been a fine breakdown of marks on a right/wrong scale. This is coupled with a definitive acceptable answer.

The two patterns, illustrated by the examples, highlight differences in philosophy to the stated objective of GCSE-CDT that 'candidates should have the opportunity to demonstrate what they know, understand and can do'.

There are also quite marked differences in the way that groups have considered the needs of candidates, whatever their ability, in being able to fully understand requirements of questions, and also the supportive structure for formulating a response. The language used, the layout for each page, the format for a response and even the type of illustration have a tremendous effect on the quality of response. Some groups have given a high priority to these requirements, whilst others have yet to refine appropriate strategies.

Consider the examples (Figures 10 and 11) in respect of:

a) The language level used, and the problems the candidate may have in understanding the information given.
b) Layout and framework for response.
c) Indication of levels of difficulty and complexity, and possible rewards.

FUTURE DEVELOPMENTS IN EXAMINING TECHNIQUES

Just as there have been major changes of approach in the classroom and workshop so we must expect major changes of examining technique to take place. Some developments have been highlighted in this chapter, but with endeavours to improve upon what has already been achieved and the experience that is still to come, refinements and changes will inevitably take place for many years to come. Much of the sample material available

SPECIMEN MARKING GUIDELINES FOR PAPER 1

The marking guidelines being distributed are offered as a guide to teachers. They are not definitive but indicate in broad terms how candidates' answers will be assessed. A marking scheme used in the operational examination will have been subject to a standardisation procedure.

SECTION A   20 compulsory part-questions

1.	*(a)*	Tenon saw.	**1 mark**
	(b)	Redwood.	**1 mark**
	(c)	Any plastics laminate.	**1 mark**
	(d)	Any suitable example.	**1 mark**
	(e)	Clear, named sketch of a round headed screw.	**1 mark**
	(f)	Any section which allows access to all parts of the curve.	**1 mark**
	(g)	Mild steel.	**1 mark**
	(h)	Spot welding.	**1 mark**
	(i)	Parting off on a lathe.	**1 mark**
	(j)	Any relevant precaution.	**1 mark**
	(k)	Gearwheel C.	**1 mark**
	(l)	All three pins must be correctly labelled.	**1 mark**
	(m)	2 amps.	**1 mark**
	(n)	Compression.	**1 mark**
	(o)	The symbol represents an earth in a circuit.	**1 mark**
	(p)	Any thermosetting plastics material.	**1 mark**
	(q)	Thermoplastics can be re-formed if heated.	**1 mark**
	(r)	Welding.	**1 mark**
	(s)	Any suitable situation (e.g. illuminated signs).	**1 mark**
	(t)	Any correct sketch.	**1 mark**

total	**20 marks**

2.	*(a)*	(i)	Beech	**1 mark**
		(ii)	Any suitable varnish finish.	**1 mark**
		(iii)	For protection, or to allow the quality of the timber to be seen.	**1 mark**
	(b)	(i)	Mild steel	**1 mark**
		(ii)	Cellulose or enamel	**1 mark**
		(iii)	Spraying or dipping	**1 mark**
	(c)	(i)	Any suitable thermo-plastics material.	**1 mark**
		(ii)	Durability, or the visual quality of colourful plastics.	**1 mark**
		(iii)	Injection moulding.	**1 mark**
	(d)		In each case give credit for observations relating to the role the toy would have in the play situation, its visual stimulus, etc.	**3 marks**

▲   *Fig. 9 Specimen marking guidelines, SEG*

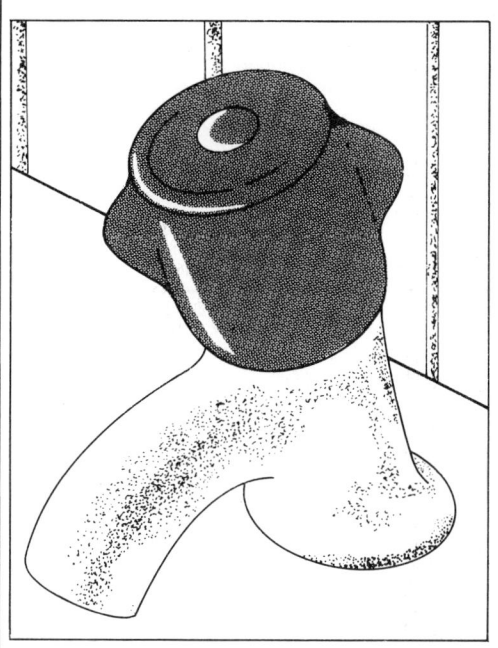

Fig. 1

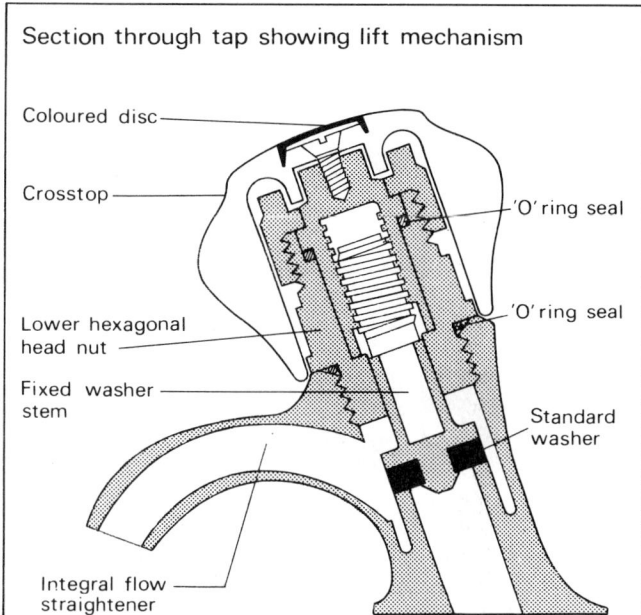

Section through tap showing lift mechanism

Coloured disc

Crosstop

Lower hexagonal head nut

Fixed washer stem

Integral flow straightener

'O' ring seal

'O' ring seal

Standard washer

Fig. 2

Figs 1 and 2 above show the "Opella" 500 Series plastic tap for use on a wash-hand basin, and a section through this tap. It was designed many years ago as an alternative to a metal tap, is still produced and is currently available in several different models.

Consider a **metal tap**.

(a) (i) List the range of materials used in the conventional metal tap. [1]

   (ii) Name the process by which the body of such a tap would be made. [1]

   (iii) Name the finish usually given to metal taps and describe clearly and with a diagrammatic layout how it would be carried out. [4]

(b) Make a concise list of the advantages that a plastic tap might have over the conventional metal model. [4]

(c) On the sectional view, Fig. 2, "O" Ring Seals are shown in two locations.

   (i) What function do they perform? [1]

   (ii) From what material are "O" ring seals usually made? [1]

(d) Also on the sectional view, Fig. 2, a coloured plastic disc is shown.

   (i) Name the process by which it would be made. [1]

   (ii) Draw clearly and carefully a sectional view of the tool used to produce the coloured disc. [3]

▲ *Fig. 10 Specimen question, SEG*

(c) The diagram shows the cross-section of a dispenser for liquids.

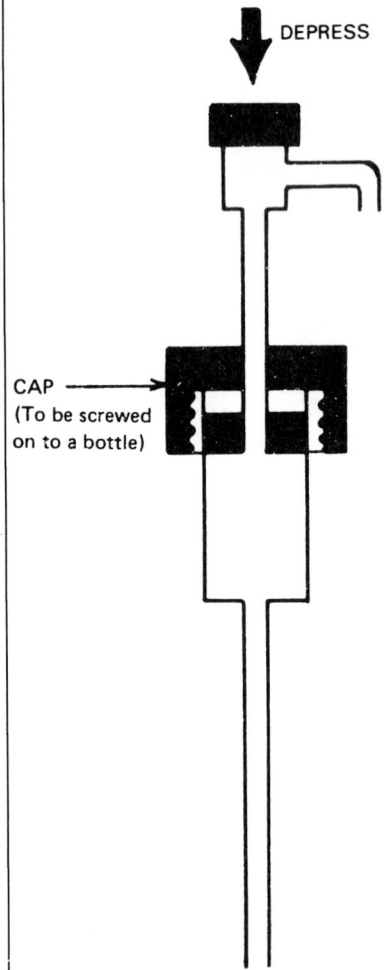

DEPRESS

CAP —►
(To be screwed
on to a bottle)

(i) Draw, on the diagram, a spring in a position which will cause the tap to return to the top of its stroke.

(ii) Mark on the diagram the positions of the two control valves.

(iii) In the space below, make sketches to show clear details of the valves.

(d) A diagram illustrating the sequence in supplying water and flushing a w.c. is shown below. Translate the diagram into a drawing(s) which show the necessary components.   [8]

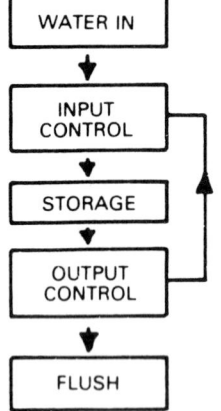

▲ *Fig. 11 Specimen question, LEAG*

Question 1

The World Wildlife Fund is to launch a publicity campaign to bring to the attention of all members of the public the need to protect all wildlife from the danger of extinction. Part of the campaign will take the form of eye-catching displays which will endeavour to explain the problem. Such displays will be mounted in shops, libraries, gardens and other such public places. So that they are inexpensive they should be made from card or similar sheet material cut, shaped and folded as required. The message they carry should be simple but make people aware of the problem. Most of all they should remind us of the beauty of living forms.

(a) Sketch three different ideas for such a display, giving an indication of size, colour and construction.   20%

(b) Choose **one** of these ideas which you consider will be most effective and give your reasons for selecting it.   10%

(c) Prepare a full-size elevation of this display.   20%

(d) Make a full-size development of this display before cutting and folding, but indicate the position of any cuts or folds using appropriate representation.   25%

(e) Using the paper and adhesive provided cut out the shape, fold and assemble.   25%

▲ *Fig. 12 Specimen question, NEA*

for illustrating types of questions and mark schemes in the documents published by the Examining Groups has been developed by committees and working parties, not the normal route taken for the production of a question paper and mark scheme, but one which had to be adopted in order to give teachers an opportunity to see what is likely to be used at least in the early stages of the GCSE. The refinements to such papers will be evident when the responsibility of setting the papers falls on the shoulders of Chief Examiners and they gain the experience of each year of the examination. Some Examining Groups are using the combined examination of GCE and CSE in 1987 as a means of gaining experience. Papers have been set and many schools have indicated their intention to enter pupils for this examination. So by the end of the examining period for 1987 past papers will be available that will be as close as possible in likeness to those that may be expected in 1988.

The major changes are now evident and such radical changes are not likely to occur again for some while. The differences between one Examining Group's syllabus, mode of assessment, weighting of elements and overall approach and that of another is so small that the basis for selection may rest solely upon the geographical region of the school.

1. (a) The drawings below show
   (i) a designer's final arrangement of components in the timer;
   (ii) the overall size of the space taken up by the components.

   Sketch at least three designs for a casing to house the components to make a timer.

   The control buttons have to be covered so that they cannot be altered accidently.

   Remember that it should be easy to change the battery.

   Your sketches should be freehand and you should make brief notes explaining your designs.

   (30 marks)

Sketches which show the final arrangement of components in the timer

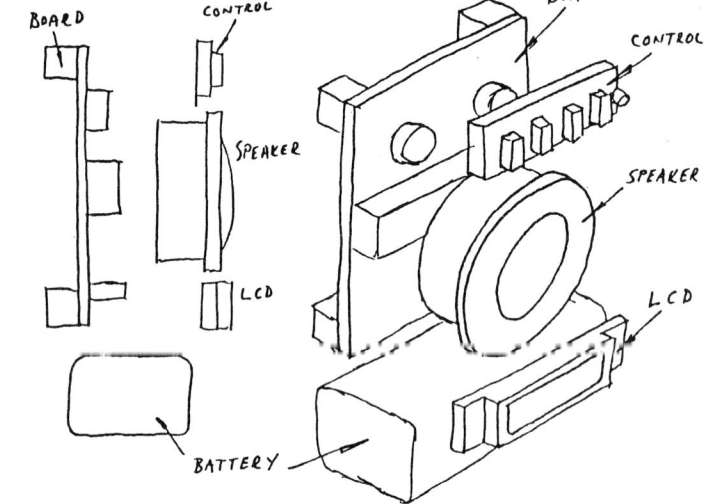

Overall size of space taken up by assembled components

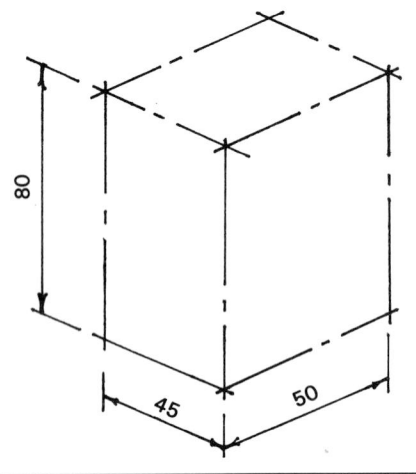

▲ *Fig. 13 Specimen question, SEG Design and Communication*

3 Project Work and Resources

Development in the three strands of CDT has been taking place during the last decade. Though perhaps the titles have been slightly modified, the underlying theme of problem-solving can be identified in many of the GCE and CSE syllabuses. Even in Graphic Communication problem-solving has been an activity in project work and compulsory assignments which contribute to assessable material. Because it is a relatively recent innovation, it is perhaps not well known and deserves special attention later in this chapter.

PROBLEM-SOLVING

Identification of a Problem

Problem-solving, as an activity, must have a starting point. The problem must be clearly identified before the whole process of design can begin. Because problems are explicitly recognised as a result of experience, it usually falls upon the teacher to do the identifying. Most pupils, if coldly asked to identify a problem, come up with an answer which is a solution to a problem. 'I want to make a small table for my mother', 'I want to design a stand for my loudspeakers' are the cries often heard by the teacher who has not yet got his pupils thinking on the right lines. 'But the table is a solution to a problem,' the teacher may retort. 'Yes, sir, but I still want to make a table.' This rather inane dialogue often leads nowhere, especially if the foundation course for CDT has been about only designing artefacts without due consideration of the problems that they resolve. Children, particularly those in the lower-ability ranges, need considerable help with this aspect of design, and there is a straightforward way of overcoming this difficulty.

Firstly, start with any solution, and then try to see what problem it satisfies. The solution may be a desk lamp, a telephone shelf, or on a larger scale a road bridge or any form of transport from a bicycle to a supersonic aeroplane. All these artefacts are solutions to problems, therefore the task is to put into words what would eventually lead to the production of such a solution. Starting with the desk lamp, the problem was that a source of lighting was *needed* to illuminate a desk sufficiently for a student to be able to complete his/her homework in the evenings. As for the telephone shelf, this may have been the solution to the *need* to accommodate a telephone, a telephone directory and a writing pad in the hall or passageway near the entrance of a house. The solution is a response to a need. The *need* for a bicycle is to provide a cheap form of transport. The *need* for a supersonic aircraft is to cover large distances quickly.

Having recognised needs, it is possible to move on to specifications or analysis of the problem.

Analysis of the Problem

In order for a designer to begin attempting to resolve the problem, it is essential to have as much information about any special requirements as possible. To go back to the desk lighting problem, it would be helpful to know whether or not the solution must be adjustable, or fixed to a wall or desk top, or can it be free-standing? Does the switch for the light have to be operated at the power source or must it be incorporated in the design?

Further analysis which has a bearing on the design will include such aspects as economy. Who will be realising the problem – an individual pupil or a team of pupils? This may have considerable influence on the form the final solution takes and the techniques that are used. Does it have to be made solely with the range of equipment that is available and the materials that are in stock? Does the making have to be completed by a particular date?

Having established the special requirements it is important to investigate other relevant information, sometimes referred to as research.

Investigation or Research

Continuing with the lighting problem, it is important to know something about the electrical fittings. Since sizes are fairly standard, these will dictate many of the dimensions in the final design. Thus it is important to establish these, including the size and type of bulb to be used. In a true research situation it would be more helpful to establish facts by trying out different forms of lighting, rather than depend on information just obtained from books. The terms primary and secondary sources are not ones which children should necessarily use, but they should be aware that personal investigation is one method of establishing facts, and reading from printed material is a second means of gaining information. The latter, however, may be less dependable than primary research, particularly if its source is doubtful.

The investigations may lead into other aspects, such as the properties of the materials, the distribution of weight and movement to achieve balance; the forces that may be created and need to be counteracted. In a project of this kind, due consideration must be given to the safety aspects, and the information should only be obtained using reliable battery-powered circuits or other secondary resources. In no way should a pupil be allowed to experiment with lighting that comes directly from a mains supply.

Development of Ideas

Ideas do not come easily. Generating them is often a period of considerable anguish when the mind either seems fixed on a single idea or is void of any ideas at all. In such circumstances some guiding remarks or reference to other responses to a similar problem may prove helpful in prompting ideas.

Once a number of ideas begin to emerge, the next task is to choose a method by which these abstract ideas may become evident in more visible and tangible forms. Drawing is perhaps the most common method used to develop ideas since it is manageable, and the materials – pencil and paper – are readily available. But it is not the easiest thing in the world to express three-dimensional forms on a two-dimensional plane. The graphical skills of most fifteen- and sixteen-year-old pupils are insufficiently developed to allow a freedom of expression that will do justice to the ideas they are attempting to express. The three or four drawings which purport to being the development of ideas are often the results of sketches drawn after the solution has been realised.

Examples of maquettes

Even such a great artist/sculptor as Henry Moore recognised this factor and developed his ideas for sculptures in Plasticine or clay. The miniature models or maquettes gave a three-dimensional image that enabled it to be viewed from all angles. Clay, Plasticine, card, balsa wood and computers, if available, are all valuable aids to developing ideas.

Trying out ideas

Empirical research has its place in the attempt to establish three-dimensional forms. Offcuts of wood arranged to simulate a variety of forms help to establish a structure or form much more readily and easily than the outlines pencilled on a piece of paper. However, from the temporary nature of the arrangements, it is important that a more permanent record is made if ideas are going to be developed, and drawing these would be acceptable and easy for the pupil. Photographing is another means of recording information of this kind, but its cost and feasibility would have to be considered and weighed against the benefits gained. A sixth-form student studying A level Design and Technology, however, could well expect to use such techniques.

Where shape and form are not the prime consideration, but rather working out the principles of levers, pulleys and gears, construction kits afford the best means of trying out ideas before any attempt is made to produce a final product.

Selection of an Idea

Though pupils, either through impatience or a true conviction that they have a good idea, may resist the suggestion to try a number of other possible solutions, it is important that they are encouraged to look at other possibilities. Without this, the single solution cannot be compared and judged on its merits. Should arguments not be convincing enough for the pupil to attempt alternative solutions, he/she may be willing to look at minor details to see what other possibilities may emerge, e.g. if the need is to produce a lighting unit that has to be fixed to a desk top, then this aspect can be developed by going back to the investigation stage and looking at different methods of fixing that are employed in other artefacts. Most drawing rooms have an office-type pencil sharpener that is fixed to a surface; would not the principle employed for the pencil sharpener be readily adaptable for fixing the desk lighting unit? Drawing attention to detail of this kind will help not only to encourage the pupil to look at alternative solutions but also to become more aware of his/her surroundings.

In order to evaluate the ideas that have been generated through investigation, awareness of similar products and maybe an injection of original thinking, it is necessary to return to the beginning – the need and the analysis. How well do the ideas, which may still only be expressed in line drawings, satisfy the requirements of the problem? Making a selection on such slender evidence is often more an act of faith than a critical analysis of all aspects of the potential design. This is where the mock-up or model may help to produce more concrete evidence, and should be the next stage.

'Mock-up' or 'Model'

Many of the ideas have been formulated can only be assessed by transforming the two-dimensional drawing into a three-dimensional form, using soft materials which are easy and quick to cut, bend and shape. The model is not expected to be made to the same standards of craftsmanship as the final article. Therefore, temporary and quick fixing methods are preferred to the more permanent and elaborate methods used for the final product. The model occupies a similar position to the sketch for a final drawing or painting. The model is in fact a three-dimensional 'sketch' made from three-dimensional materials.

The functional details may be examined more closely on a model, particularly if there are moving parts. An adjustable lighting unit, which may have to be capable of adopting a variety of positions, will not work if one part of the structure gets in the way of another moving part. Adjustment of sizes on any of the component parts may be all that is necessary to allow the range of movements that is required. Finding these out on a model is much cheaper and less time-wasting than on the finished product.

Having established that the design is capable of achieving all the movement functions efficiently, it is necessary to put the information down in sufficient detail for it to be realised in a resistant material, i.e. wood, metal, plastic, etc.

The Working Drawing

For many people, this means a drawing in Orthographic projection, and for a number of situations this type of working drawing is ideally suited. However, it would be wrong to conclude that all working drawings should be drawn in Ortho-graphic, first- or third-angle projection. Abstract shapes in jewellery design or free-standing sculptures are very difficult to present in the more formal ways that suit a piece of engineering or cabinet drawing.

A working drawing is first and foremost a drawing from which the artefact can be made. It contains all the details drawn to scale. Where possible the scale should be one:one or full size, i.e. the same as the finished item. Then it is possible to use the drawing as a means of checking that all components agree with the drawing by placing each part on the actual drawing. By using first- or third-angle projection, most surfaces will be 'true shapes', but inclined planes will need to be drawn with an auxiliary projection, i.e. one that is drawn on a plane parallel to the sloping plane, in order to achieve a true shape.

To judge whether or not a working drawing is complete, the question must be asked, can the artefact be made, from the details given, by someone other than the person who developed the ideas and produced the drawing? A good working drawing should at least:

a) be drawn to scale, preferably full size;
b) contain all the necessary details about construction;
c) give details of materials to be used;
d) contain a parts list;

e) state the 'finish' that is required;
f) conform to 'recommended British Standards'.

These drawings should not be coloured, but may be produced in pencil or black indian ink. Print-outs from computers are of course also acceptable, provided they are the result of the pupil's own programming.

Realisation

This is the term most commonly used to describe making the final artefact, since it forms an integral part of the designing activity. This is the area of work which in the earlier years of Craft Education occupied as much as 80 per cent of the teacher's time, and, of course, the pupil's. Now the activity of making the artefact is reduced to 40 per cent, and in the same way it contributes only 40 per cent of the marks assessable. This change of emphasis has not come without being criticised, but the criticism has come mainly from experienced craft teachers whose enjoyment of craftsmanship left very little time for anything else.

However, the reduction of time for realisation does not in any way suggest that its status and importance is less respected. The success of any artefact must ultimately depend upon the quality of its production.

In order that a solution to a design problem may be realised according to the designer's intention, more than a single type of material will normally need to be used. Wooden components may have to be joined with metal or plastic components, so opening up a whole new range of joining processes. The techniques of joining wood to wood and metal to metal have evolved over many hundreds of years and so have resulted in highly sophisticated methods of joining; the dovetail is one such joint. Its proportions and precision of angles depending on the type of wood being used are as much a result of individual taste as they are of proven testing. Whereas in practical examinations, certainly, the dovetail formed a compulsory assessable component, this is no longer the case. A pupil need never make a dovetail from the CDT foundation course in the first year of secondary education to the time of completing a CDT course at the end of the fifth year. The skill of making still requires the same degree of accuracy in measurement and exactness of fit as it has traditionally, but the method of joining is selected essentially as a result of decisions made during the designing stage.

Length of material, squareness of edges and overall quality of finish are all that is necessary to assess the degree of competence achieved in working with resistant materials. When faced with the task of assessing quality of craftsmanship, the design aspects must be ignored. These are assessed separately under the heading of 'Suitability of solution to fulfil the requirements', in which case one judgement has been made and a second judgement of the same aspects cannot be made under another heading. It is quite as reasonable to expect to find examples of poor designs made very well as it is to find good designs made badly.

With the introduction and the availability of manufactured boards, advanced bonding techniques for wood, metal and plastics, computer-aided machines and the like, differences in design solutions must inevitably be wide-ranging. Gone are the days when every pupil in the class would be involved in the production of an identical artefact, and the only difference in the activity of each pupil would depend on the rate at which he or she could work. The implication of this is that practical rooms and organisation of tools, equipment, materials, etc. will have to change considerably to cope with the demands made by CDT courses.

Evaluation

Though this has been a component of many previous examination courses, it is now a part of all courses that have a CDT prefix.

In some aspects the product may be such that experimental tests may be carried out on it to establish just how suitable the design really is, in which case the evaluation may take the form of a report. Results will be tabulated and objective judgements made and recorded on a standard report form. The report form will be one that the school has devised to suit the type of work being carried out.

Judgements made about artefacts which are not suitable for being experimented with will be made using less formal observations, and the comments will be much more subjective. Nonetheless the level to which an evaluation can be made should be quite critical and show a depth of understanding of the features that gain merit, as well as weaknesses. Where a weakness has been observed, a good evaluation will contain recommendations for improvement. If the end product is to perform certain functions, then these can be assessed quite objectively. Is the lighting unit easily adjustable? Does it remain steady in all its adjustable positions? Is the point of fixture on the desk adequate? Does it hold the unit steady even when adjustments are being made? Is the luminosity of the right intensity? Does the unit in any way impede working on the desk top? In other words, does it fulfil the identified problem and analysis?

With the evaluation completed, the pupil will have experienced many of the tasks that a professional designer faces today. To illustrate what a pupil may be involved in, a problem related to lighting was chosen, but the problem could have been related to a situation in which resistant materials such as wood, metal and plastics are not required.

PROBLEM-SOLVING IN DESIGN AND COMMUNICATION

Many of the Graphics courses that are available at GCE and CSE contain an element that involves the candidates in producing a folder of information on a given topic so that they may be able to answer questions in a formal two-hour examination. Those teachers involved with Graphical Communication offered by AEB will be familiar with what is required, and should not find the demands of CDT Design and Communication so very different.

Problem-solving already exists in many of the GCE and CSE Mode 1 examinations. Though there will be differences in the requirements and methods of assessment used, it will be helpful to look at what may be a representative set of examples, rather than a number of specific examples taken from a single document.

General Pattern of Organisation in GCE and CSE

All students will require to be formally examined on the aspects of Geometry, plain and solid and projection techniques. This is quite standard for many Technical Drawing subjects that have been in existence for more than half a century, but more recently Design has been included and has become the major, integral part of the syllabus. 'Design in relation to graphical presentation will be an integral part of all sections of the examination.' (University of London GCE O level Graphical Communication). This means that many of the questions are structured in such a way that they become 'mini problems' to solve. The candidates are required to analyse given problems, to show an understanding of the problem and to present a solution graphically in answer to the given information. Figure 1 shows an example where a graphical solution is required.

---

9. The labour force for railway maintenance in a Third World country included people of several different nationalities, not all of whom can read or speak the native tongue of the country. It is necessary therefore that a number of pictographic informational signs be posted at various points to warn of potential hazards.

The main body of information is as follows:

Weak roof	(Do not walk on it unless supported on ladders and/or planks)
Weak wall	(Do not lean ladders against it)
Electric cable	(Do not touch or pierce)
Gas pipe	(Do not heat or pierce)
Water mains	(Do not pierce)
Underground pipes and cables	(Do not dig here)

Select three from the above and produce on the plain A2 sheet provided, a number of design sketches for each. Take one of the sketches of one of the designs to a finished solution.

Note: Since a selection of the pictograms may be required on a single sign, a uniform style for each is important. Marks will be awarded for the appropriate use of colour. **(20 marks)**

▲ *Fig. 1   ULSEB, June 1985*

Here the candidate is presented with a design problem which has to be completed under examination conditions. The candidate has to answer four questions in the space of three hours without access to any reference material.

A list of study areas is given at the beginning of the course, or can be, if the teacher so wishes, passed on to the pupils, because a list is printed with the syllabus and is very extensive. In June 1985 the study areas included:

a) School creative subjects;
b) Manufactured articles commonly found within the school or home environment;
c) Logos, symbols, ideograms;
d) Line graphs, bar charts, pie charts and mathematical patterns;
e) Diagrams, time charts, flow diagrams to illustrate scientific historical/geographical data (BS 4058);
f) Instruction sequences for do-it-yourself and model making;
g) Maps, charts, layouts of buildings and contents rooms, roads (BS 1192);
h) Elementary electrical circuits (BS 3939);
i) The construction of static and working models.

Faced with this list the pupil had the choice of studying all areas or concentrating on selected ones, knowing that the question paper carried a range of choice questions as well as compulsory ones. Question 9 obviously suited those pupils who studied 'Logos, symbols, ideograms'.

Another GCE examination board, AEB, adopted a different approach in that a Design study topic would be announced approximately twelve weeks before the date of the examination. The topics over the last five years have been based on the following:

Adjustable lighting
Scissor action jack
Garden pressure sprays
Torches etc.
Cantilever tool boxes
Tubular furniture

The candidates are required to compile appropriate information through their own investigation and present it in a design folio. It was expected that the folio not only contained the information surrounding the topic, but also that it would provide the pupil with the opportunity to demonstrate a variety of graphical skills ranging from sketching to two-point perspective drawings rendered in coloured inks, crayons, etc. and more recently airbrush techniques. This of course was done in the pupils' own time and not within the constraints of a formal examination. The final product – the design folio – also had a weighting of 15 per cent of the total marks and so presented an occasion where coursework formed an element of the assessable material. The work was assessed by the pupil's own teacher, who followed an assessment scheme that was supplied by the examining body and moderated by one of the Board's appointed Moderators.

One of the examination papers that followed in June presented a series of questions that aimed to determine the

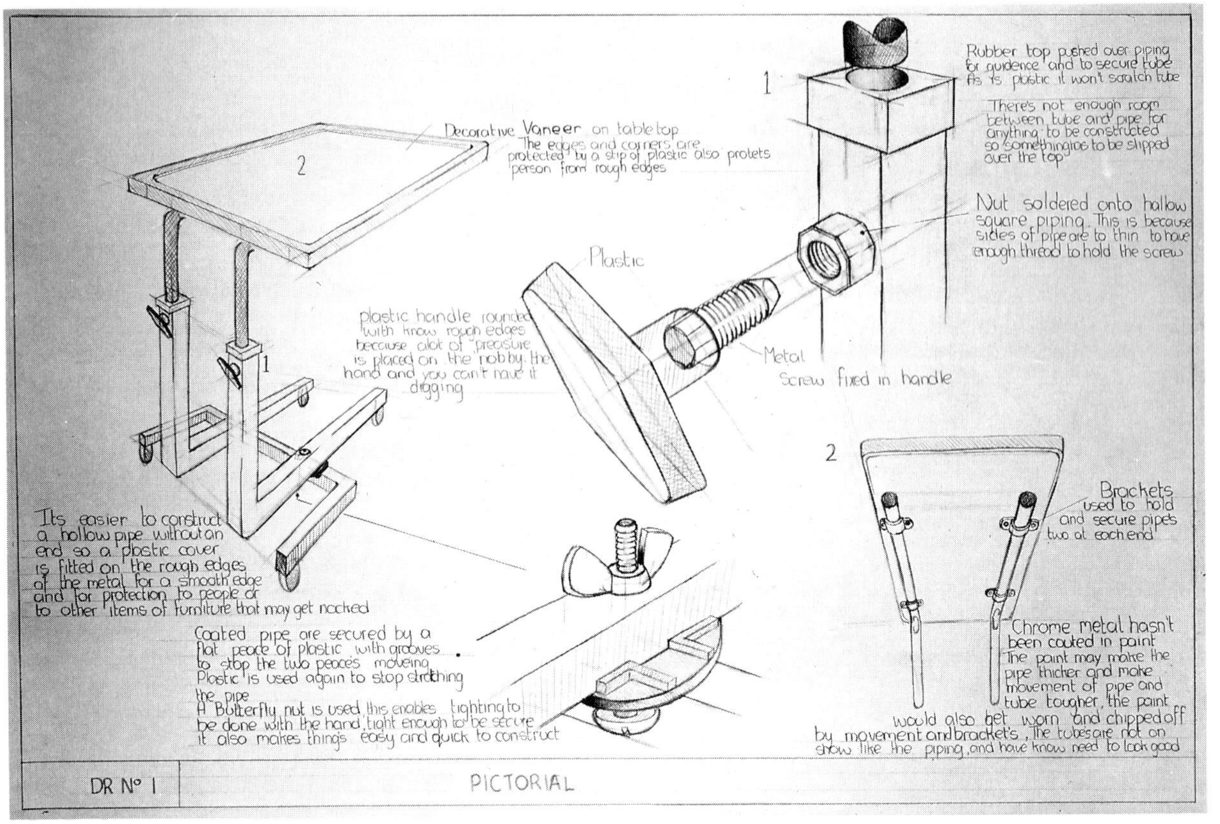

DR No 1 — PICTORIAL

Decorative Vaneer on table top
The edges and corners are protected by a strip of plastic also protets person from rough edges

1

Rubber top pushed over piping for gurdence and to secure tube as its plastic it won't scratch tube

There's not enough room between tube and pipe for anything to be constructed so something has to be slipped over the top

Nut soldered onto hollow square piping. This is because sides of pipe are to thin to have enough thread to hold the screw

Plastic

plastic handle rounded with know rough edges because alot of pressure is placed on the nob by the hand and you can't have it digging

Metal
Screw fixed in handle

Its easier to construct a hollow pipe without an end and so a plastic cover is fitted on the rough edges of the metal for a smooth edge and for protection to people or to other items of furniture that may get nocked

Coated pipe are secured by a flat peace of plastic with grooves to stop the two peaces moveing Plastic is used again to stop stretching the pipe
A Butterfly nut is used this enables tighting to be done with the hand, tight enough to be secure it also makes things easy and quick to construct

2

Brackets used to hold and secure pipes two at each end

Chrome metal hasn't been coated in paint The paint may make the pipe thicker and more movement of pipe and tube tougher, the paint would also get worn and chipped off by movement and brackets. The tubes are not on show like the piping, and have know need to look good

---

DR No 3 — Ways in which two panels are joined together

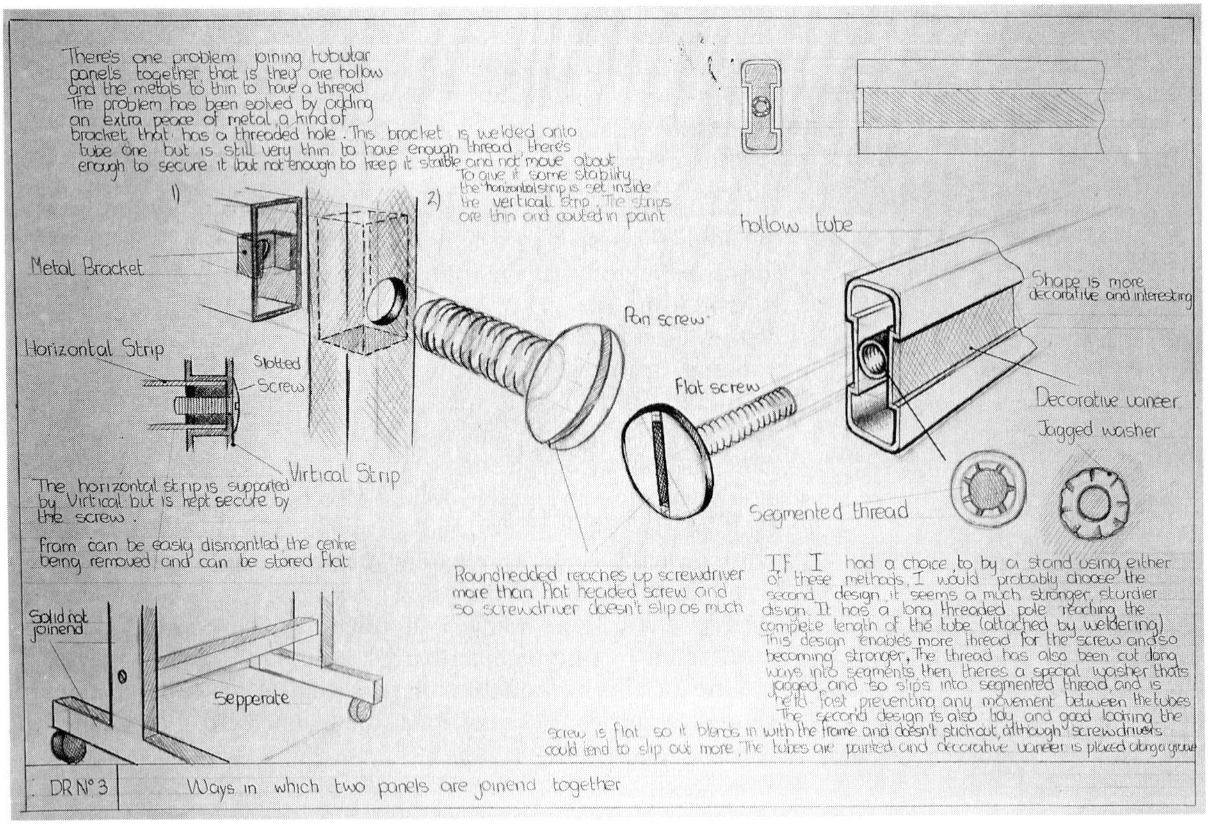

There's one problem joining tabular panels together that is they are hollow and the metals to thin to have a thread The problem has been solved by adding an extra peace of metal a kind of bracket that has a threaded hole. This bracket is welded onto tube one but is still very thin to have enough thread, there's enough to secure it but not enough to keep it stable and not move about
To give it some stability the horizontal strip is set inside the vertical strip. The strips are thin and coated in paint

1)

2)

Metal Bracket

Horizontal Strip

Slotted Screw

Virtical Strip

The horizontal strip is supported by Virtical but is kept secure by the screw.

Form can be easily dismantled the centre being removed and can be stored flat

Solid not joined end

Sepperate

Round headed reaches up screwdriver more than flat headed screw and so screwdriver doesn't slip as much

Pan screw

Flat screw

hollow tube

Shape is more decorative and interesting

Decorative vaneer

Jagged washer

Segmented thread

If I had a choice to by a stand using either of these methods, I would probably choose the second design it seems a much stronger, sturdier design. It has a long threaded pole reaching the complete length of the tube (attached by weldering) This design enables more thread for the screw and so becoming stronger, the thread has also been cut long ways into segments then theres a special washer thats jagged and so slips into segmented thread and is held fast preventing any movement between the tubes The second design is also tidy and good looking the screw is flat so it blends in with the frame and doesn't stick out, although screwdrivers could tend to slip out more. The tubes are painted and decorative vaneer is placed along a groove

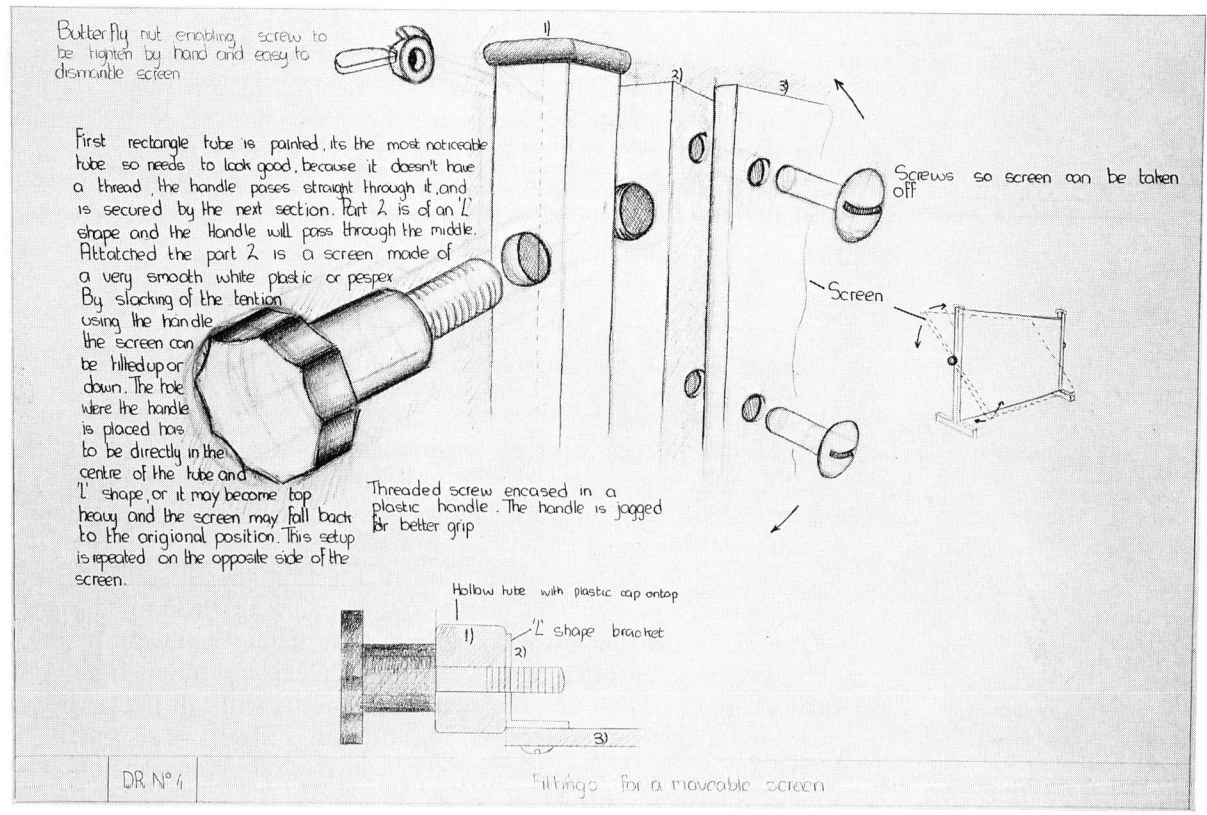

Butterfly nut enabling screw to be tighten by hand and easy to dismantle screen

First rectangle tube is painted, its the most noticeable tube so needs to look good, because it doesn't have a thread, the handle poses straight through it, and is secured by the next section. Part 2 is of an 'L' shape and the Handle will pass through the middle. Attatched the part 2 is a screen made of a very smooth white plastic or pespex. By slacking of the tention using the handle the screen can be tilted up or down. The hole where the handle is placed has to be directly in the centre of the tube and 'L' shape, or it may become top heavy and the screen may fall back to the origional position. This setup is repeated on the opposite side of the screen.

Screws so screen can be taken off

Screen

Threaded screw encased in a plastic handle. The handle is jagged for better grip

Hollow tube with plastic cap ontop

'L' shape bracket

DR N° 4   Fittings for a moveable screen

candidate's ability to analyse problems and to present solutions. The design folio, however, could not be taken into the examination, and therefore the candidate had to rely on what could be remembered from the study.

Some CSE Mode 3 schemes went considerably further on both the amount of coursework that could contribute towards assessment and implementing a more complete Design approach. The designing element and all its associated activities are comparable with those of a Design and Realisation course. The elements of identifying a problem, analysing it and presenting a concrete solution in either card, balsa wood or any other suitable material through to an evaluation, could be readily identified in the project coursework. This is the direction in which many of the GCSE Graphic Communication courses and CDT Design and Communication have moved, but a close examination of Chapter 2 will help to highlight the differences between them.

GCSE DESIGN AND COMMUNICATION

For purpose of reference, the Southern Examination Group's Syllabus and Assessment Pattern has been chosen to illustrate this section. It follows on nicely from the developments and changes that have been taking place in GCE and CSE design content, and helps to show the progression that has taken place.

CDT Design and Communication has its roots in two sources, Crafts and Technical Drawing. After these were combined with the developments that have taken place in design education, the problem-solving activity became the major feature in all the GCSE Design and Communication courses that have been approved by the SEC. This can be seen very clearly in the SEG pattern of assessment:

Paper 1 based on knowledge carries 20 per cent of the marks.
Paper 2 based on design carries 30 per cent of the marks.
Paper 3 coursework based on problem-solving carries 50 per cent of the marks.

Hence 80 per cent of the total marks are available for the design element. This clearly indicates the importance given to the skills of designing and shows the direction in which teachers must guide their pupils.

It is expected that at the end of the first year of the course pupils will be knowledgeable and capable of tackling design problems, so that during the final year many of the problem-solving activities can be done at varying levels of independence. It should be remembered that some pupils will not necessarily follow a design process rigidly and will still arrive at a viable solution. A little flexibility here may be desirable for some pupils and for certain projects.

Assessment of Coursework

The project for the final year is determined by the pupil in discussion with the teacher. It is envisaged that the more able pupils will be able to identify a problem and suggest ways in which the problem may be tackled. A clearly detailed statement should be the hallmark of a good candidate, because all too often with such an open-ended situation the teacher has to provide a number of possible 'design statements' from which either the less able or poorly motivated have to choose. Otherwise time will be wasted and disciplinary problems arise.

Because credit is given to the pupil or pupils who initiate their own problems and design briefs, it is important for the teacher to keep a record. All examining groups have a standard coursework assessment form which itemises the designing aspect into at least five main areas: identification of problem, analysis, development, evaluation and making. Marks are awarded for each aspect and can be recorded by the teacher as and when each aspect is completed. Under 'identification of problem' the mark awarded must reflect the teacher's involvement, if any, so that the pupil who produced his/her own detailed design brief without assistance scores more highly than the pupil who was supplied with a design brief.

To enable the teacher to assess the work, sample guidelines have been given (see pages 71, 76, 84 and 86). It must be remembered that Examination Groups may wish to modify these marking guidelines, and teachers should check that the guidelines they are using are those applicable to the coursework that is being assessed. Many teachers who are responsible for marking coursework have, in the past, failed to recognise that changes are made periodically and new marking schemes and guidelines are published. While some have argued that the Examining Group is at fault for not sending the revised scheme, it has often been discovered that the information has been sent to the centre and the subject teacher has not checked, even as a precautionary measure, to see if changes have been made. Faults of this kind can occur on both sides and, with so much change taking place, even more stringent efforts are needed to ensure smooth running efficiency.

Authentication

Examiners have to abide by very stringent regulations when marking scripts in order to ensure that scripts are marked thoroughly and properly. They also have to sign a document agreeing to the conditions that have been laid down by the Group. Now that coursework plays such an important role in providing assessable material, teachers responsible for carrying out the assessment must be prepared to accept conditions that are stipulated by the Examining Groups. Printed on the bottom of a specimen Coursework Marksheet is the following:

Declaration by teacher

I confirm that the work of the candidates listed above has been kept under supervision and that, to the best of my knowledge, no assistance has been given apart from that which is acceptable and has been identified and recorded on the work.

Name of teacher ....
(capital letters)

Signature ....

Date ....

But another innovation is that the candidate must also make a similar declaration:

Declaration by candidate

I confirm that I have completed my coursework without any assistance except where shown.

Signature ....

Date ....

Because work can be done at home as well as at school, it is impossible for the teacher to know exactly what assistance has been given at home, if any, and only the candidate can possibly know what help has been given. A great deal is based on trust, but if a teacher is in any doubt about work that he/she has not been in a position to observe, then the Examination Group responsible for conducting the assessment must be advised, so that appropriate action can be taken. These measures may sound a little frightening at first, but they are only there to ensure that the final awards are fair and their credibility is unquestioned.

Moderating Coursework

For subjects like the AEB Craft Design Wood, Craft Design Metal, Design Communication and Application and more recently the AEB's CDT Design and Realisation which is really a replacement for the other three, coursework has been assessed by the teacher. There are, of course, many CSE subjects in which teacher assessment is necessary and therefore those teachers who have been involved will know what to expect. However, CDT Design and Communication is a new development, and only limited experience of coursework assessment has been gained, so it is with those teachers in mind that this section is included.

Once all the work has been assessed in a school or centre there is a need for moderation to take place.

Internal Moderation

The first stage of moderation must take place in the school, particularly where there is a large entry of candidates and more than one teacher was involved in the preparation and assessment of the candidates' work. Whatever practice is adopted, all the teachers involved in the assessment must be included in the first stage of moderation. There are two good reasons why this is necessary.

1. The teacher making the assessment should be the pupil's teacher, for only he or she can know to what extent the work presented is the result of the pupil's own aided or unaided efforts.

2. Cross-moderation can only take place effectively if those who are personally involved have the opportunity to justify the marks that have been initially awarded. A change of mark or revision should only be made with the agreement of all concerned.

This first stage of moderation is crucial if the next stage is to be applied effectively
Of course in the case where only a few candidates are involved and one teacher is responsible for the preparation of pupils and the assessment of the coursework, cross-moderation is not possible. The only thing that could be of assistance is in

asking a colleague who knows some of the pupils, and who is involved in a similar situation of assessing coursework, to make broad professional observations about what is exhibited. Through discussing the expectations of certain pupils – even in different disciplines – a pattern may emerge which confirms your judgement or suggests that a closer look is still necessary. If nothing else happens, the experience of discussing aspects of the work with a colleague should help to clear your thinking and be a rehearsal of and preparation for the external moderation that will soon take place.

External Moderation

The scale upon which external moderation is going to take place means that an army of Moderators will need to be employed if every school, or centre, is going to be visited. The enrolment will include many practising teachers so that every school in the country will have on its staff teacher moderators. This may prove to be a most positive move, since it ensures that there is someone who is familiar with general procedures of moderation to be available for consultation on every staff roll.

The external Moderator is the person responsible for ensuring that the schools he/she visits compare equally with the standard that has been agreed by the Examination Group. This means that all external Moderators will have attended a 'standardisation meeting' in which the Chief Examiner or Senior Moderator has established the acceptable performances for each of the grades. The procedures will vary slightly from one Examination Group to another, but the objectives will all be alike, i.e. to identify responses that agree to national standards and to ensure that the task of moderation is conducted with the highest degree of efficiency possible.

Preparation for the visit

Clear instructions should be available for every subject, and the teacher is advised to read and follow the instructions very carefully. A date and time has to be agreed between the Moderator and the school. Dates and times should be confirmed, because a Moderator may have many miles to travel and several visits to make. Once an agreement has been settled it should only be changed when unforeseen problems occur that make it impossible for moderation to be conducted.

Presentation of coursework

Again, close attention must be paid to the detailed instructions issued by the Examining Group. The work of all pupils must be displayed, or a representative sample covering the full range must be displayed and the remaining work must be available should it be required. The Moderator usually wants to browse through the work in his/her own time without the presence of anyone else. It is important, therefore, for the work to be displayed in a room that is removed from any possible distractions or disturbance. A chair and desk in the room is a consideration that is appreciated by all Moderators.

The candidate may have to be the person responsible for arranging his/her own coursework, so this point is worth checking. Also the candidates may have to select what they believe will do them justice if only a sample is required. This responsibility is often rewarded within the assessment scheme, but only in procedures that insist on every candidate presenting work. The space or area to be taken up by the coursework will be determined by the teacher who is following instructions that give details about the order that is required. So long as the moderator can be positive about whose the work is, see clearly the boundary of each candidate's work and has the teacher's marks ready to hand, his/her task of moderating should be straightforward.

While moderation is taking place the teacher responsible for arranging the display should be available if required. The other teachers who were involved in the internal assessing and moderating should not normally be required and need not make arrangements to be available.

Quality of presentation is always an important factor, since it helps to establish first impressions and thus help to put the Moderator in a good or bad frame of mind. That is not to suggest that very well presented work will automatically receive higher marks than poorly presented work. Many other qualities are looked for before such judgements are made.

The presentation of work does not begin on the day that the work is presented for assessment. It is more likely to have started at the same time as the project. In which case the orderliness, the care taken, the choice of the materials used, etc. will reflect something of the character of the candidate.

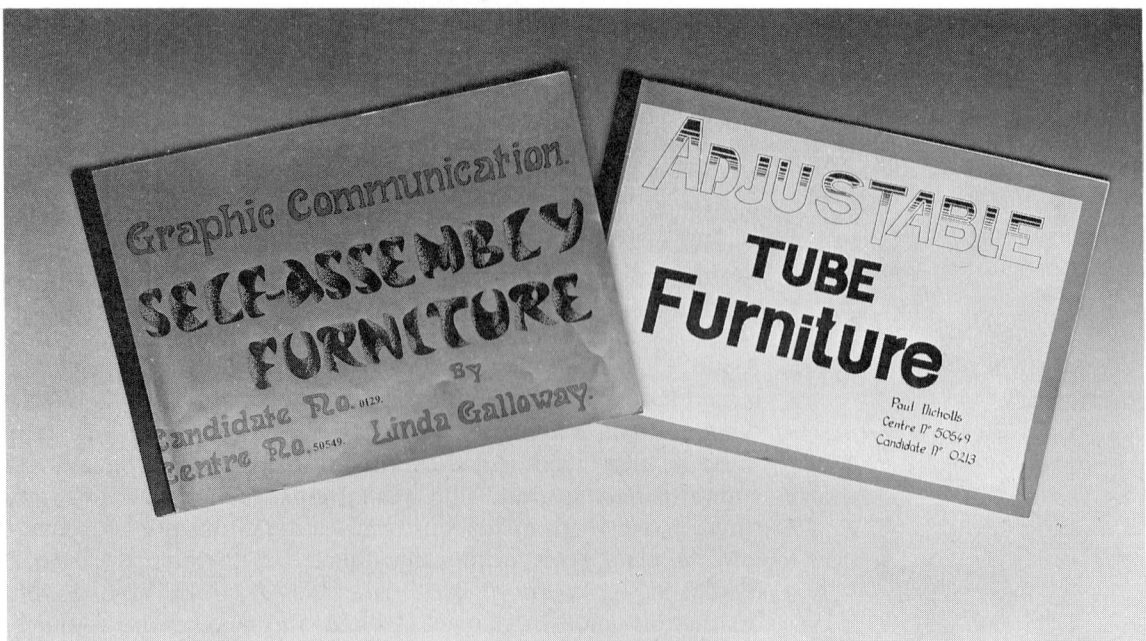

Design folios

Presentation of a Design Folio

It is not essential for a lot of expense to be incurred in the production of a folio. Purchasing ring-binder folders of A2 and A3 size can prove expensive. All that is needed for a cheaper folio is a stiff piece of card for the back, a piece of thin coloured card for the front and a plastic binder clip to hold the front and back together, allowing sheets of cartridge paper to be easily inserted as and when necessary. The size of paper used is often left to the discretion of the candidate and teacher. A2, A3 and A4 are all acceptable sizes, but A3 is perhaps the most useful. The top cover should display only the essential details, i.e. pupils's name, title of project and school, with a space left for the candidate number to be filled in when known. It is not anticipated that a candidate number will be known when the project is started. The choice of lettering, colouring and arrangement can be left entirely to the pupil's own imagination. The top cover is required to perform the same function as the cover of a book, i.e. communicate essential information in an attractive and pleasing manner. The folio can also adopt other features of a book – each page or sheet numbered, list of contents, introduction, development, conclusion, etc. will all help to present the work in an orderly manner. Whatever size of paper is selected, it is important that the folio contents are never folded.

Realisation or Making

It is possible that the contents of a folio could fulfil the realisation as well as the design activity that resulted in its production. If the problem was related to that of designing the logo for a company, the end product will be a carefully presented illustration. It should reveal all the manufacturing qualities that a three-dimensional artefact could display, in that measurements are given and precise techniques of penmanship, brushwork, airbrushwork are used. If the logo is framed and mounted, such practical skills as cutting vertical and bevelled edges can all be displayed and be comparable with the practical skills involved in producing a three-dimensional model in wood, metal or plastic.

The designing and making of pop-up illustrations for children's books may also be suitably presented in a design folio. So it is not essential to think that the 'make' element in Design and Communication must always result in a three-dimensional artefact.

The three-dimensional products in this section are often referred to as 'models' and must not be confused with the term 'model' in the Design and Realisation or Technology strands where it usually serves as a means of working something out which can be made at a later stage in more substantial materials. The 'model' in Design and Communication is the finished artefact and must show all the qualities of something well made. The pop-up illustration for a children's book may have been worked out at its development stage simply by cutting out pieces of paper to test movements and to assess their visual appearance, but the final design will be the result of considerable care and attention to detail and expertise of production.

Type of Project

Since this is a fairly new aspect of Communication, there is very little evidence of work done in this field, and the teacher may have to do more initiating of ideas in this strand than the other two, where examples of past work are much more readily available.

The list of topics that follow are provided to help the teacher get started, and to stimulate ideas for other projects, in addition to those already discussed.

1 Containers and packaging
2 Children's cut-out and construction toys in card
3 Display furniture for setting off jewellery in a show case
4 Demonstration models to show mechanical movements
5 Demonstration models to illustrate anthropometric data
6 Perpetual calendars
7 Layout of rooms with specific functions with moveable units
8 Layout of a development site, i.e. a recreational complex
9 Display stands for exhibitions
10 Competitive board games
11 Conservation of resources
12 Visual presentation of the production of artefact

The topics as they are written do not form a design brief. It is necessary to express the topic as a *need* so that ideas can be developed, e.g:

The need
A soap manufacturer requires a novel, attractive design for a container to market a cone-shaped tablet of soap.

Analysis
The container must be:

a) similar in form to the shape of the soap;
b) easy to remove from the soap without ripping;
c) attractive and eye-catching;
d) made from a single piece of material.

Further analysis which may be supplied by the teacher or the pupil:

a) The container must be made from materials available in stock.
b) It must be made using the facilities that are available in the room.
c) The design must be completed and evaluated by 31 May, etc.

All design problems have constraints, and it is helpful to identify these at the design brief stage. It would be quite disastrous for a pupil to embark on an idea which cannot possibly be realised because of the restrictive availability of materials or equipment. All the planning and thinking would be wasted and the enthusiasm generated in arriving at a possible solution would be dashed – which might damage the pupil's self-confidence when pursuing future projects.

The range of materials and equipment in a design studio will vary from school to school, so the important factor here is that the pupils are familiar with what can and what cannot be achieved.

Evaluating the Project

Evaluating is not something entirely new to the Craft teacher. Fitness for purpose was a common criteria: 'Will it do the job it is intended for?' However, this has now been considerably developed with the introduction of Design Education in the 1960s. 'How well does the solution fulfil the requirements of the design brief?' For the pupil to make this evaluation he/she will require the answer to many questions relating to all aspects of the solution. To identify the type of questions that need to be answered the design brief should be the source of reference. A well-worded brief with precise requirements or specifications can easily be rephrased to provide the necessary questions that need to be asked.

With reference to the problem of designing a container for a conical form of soap tablet:

a) (i) Does the form of the container match the form of the soap?
   (ii) Does the container form a good fit round the soap?
b) (i) Can the container be removed without being ripped?
   (ii) Can the container be removed easily?
   (iii) Does the container come off too easily?
c) (i) Do you think the container is attractive?
   (ii) In a survey, how many people out of 10 found the container attractive?
   (iii) Of those who found the container attractive, could they specify what it was they found attractive, i.e. colour, shape and so on?
d) (i) Was the container made from a single piece of material?
   (ii) Was the colour and thickness of the material suitable for the task it was required to perform?
   (iii) Was your development the most economical shape that could be produced?
And so on.

The answer to many of these questions could be just a straightforward 'Yes' or 'No', so it is important to tease out more information by asking 'Why?' For example:

b) (i) Can the container be removed without being ripped? Answer: 'Yes'. 'Why?' Answer: 'Because the design does not have any edges joined by glue and the tag and slot method used make it easy to remove without ripping.'

or

b) (i) Answer: 'No'. 'Why?' 'Because the material I used was too thin and it ripped very easily.'

With all aspects of the realisation being evaluated, it is necessary to seek recommendations for improvement or alternatives which might work equally well. The pupil who goes on to make worthwhile recommendations will certainly show differentiation in performance to the pupil who could only supply a few 'Yes' or 'No' answers.

The design should also be reviewed as a whole so that a conclusion can be made as to its general success in relationship to the statement of the need. For example:

Conclusion
The container does fulfil the main requirements and, although there is room for improving its appearance, it should provide the soap manufacturer with a novel container that should help to market the product.

In writing an example evaluation of this kind there is always the danger of producing a prescription for other evaluations; so it must be seen as only one way of providing a response from pupils which will tell us just how much they have gained from the experience. There are many ways of tackling this problem, and it is up to the teacher to decide how he/she will direct and encourage the pupils to elicit the information that is appropriate.

Just as the identification of a problem and the statement of the design brief can be formulated by an able pupil, so can the formulation of the evaluation. It is the middle to lower ranges of ability that may require a standardised format, presented by the teacher, to draw out what the pupils understand and know. In this case there is considerable justification for such a technique to be used. But, when it comes to assessing the pupil the teacher will know the level of his/her participation that was necessary in gaining this response.

Design in Society

This is a new element which has become an important part of CDT. It applies equally to the three strands, with considerable overlapping. As an aspect of Design and Communication it may at first seem less appropriate than as an aspect of Design and Realisation, and Technology. But all teachers of Design and Communication are preparing young people for the future, and it is appropriate at this stage to remind ourselves of the educational aims and objectives that refer to this aspect.

Aims and objectives

Aim No.7

To encourage technological awareness, foster attitudes of coop-
eration and social responsibility and develop abilities to enhance
the quality of the environment.

Aim No.8

To stimulate the exercising of value judgements of an aesthetic,
technical, economic and moral nature.

Objective No.17

Describe the interrelationship between design/technology and
the needs of society.

Making pupils aware

It is only too easy at any level of problem-solving to be
concerned with designing a product and to provide a solution
that satisfies a specific need. Once that product or system
becomes part of our lives, it is going to affect them in one way or
another. To bring this home to the pupils we could take the
topic of 'containers and packaging' further by examining more
closely the amount of material that is used to package products.
Why is it necessary to put a package round a product in the first
place? If it is to protect it in transit from manufacturer to the
wholesaler, to the retailer, to the customer, then this could
arguably be justifiable on the grounds that it is to ensure the
goods are received in good condition. However, as with the
design brief posed earlier, no reference was made to the
economic use of energy or materials, and more importance was
given to its form and appearance. So the design brief could be
adequately resolved without any concern for misuse of materials
or the effects of pollution that may occur during production.

These concepts may be beyond the reach of the lower-ability
pupils, but their attention can be drawn to the 'throwaway'
aspect of containers once these have performed their function.
Did the pupils who completed the 'container packaging' topic
realise that they were spending their time and energy on
designing something that was going to be thrown away? Did they
realise that they may, as designers of a package, be contributing
to the all too evident problem of litter?

Having solved one problem, another is created. How is the
litter removed? Who performs this task? Are people paid to
remove litter? Where does the litter go? These are the problems
that have to be resolved once a product has outlived its useful
life. Problems are created at both ends of the cycle. Many of our
packaging products started as plant life, so how long can we go
on producing paper and card products without drastically
reducing the tree population of the world to a level where soils
easily erode, weather patterns change, plants and animals
become extinct? The piece of paper thrown in the waste bin
must seem to the pupil very remote from the needs of

conservation of plant and animal life, but there awarenesses can be aroused by the skilful and enlightened teacher.

Using a 'problem' to make pupils aware

The topic 'Visual presentation of the production of an artefact' could be worded in such a way as to provide an excellent design problem.

The need
A lecturer requires a static visual display to show the life history of product X from the material resource to the scrap heap.

Analysis
1 The display should depend largely upon illustration techniques.
2 The illustrations should convey how the raw material is obtained, transformed into a marketable form, made into a product and how it is disposed at the end of its useful function.
3 Each illustration sheet is to be designed to fit on an A2 sheet.
4 The final illustrations must be designed to convey the advantages of having such a product, and to highlight the disadvantages in terms of conservation and pollution.

The product could be selected by the pupil and could include anything from a paper clip to a television set. Instead of a product, the design brief and analysis could be phrased to include the production of energy – electricity, gas, petrol, etc. All these topics raise issues of an economic and moral nature and lend themselves as educational vehicles for making children aware of the social implications that occur from development in technological terms and the human drive to change the environment.

The motives for change are often very reasonable, but the consequences of bringing about that change are often ignored or dismissed as relatively unimportant. It is important, therefore, to develop a strong sense of awareness so that the future generations can enjoy the changes that have been inherited rather than to have to face the task of repairing the damage that was caused by a generation of blinkered, short-sighted individuals.

ROOM MANAGEMENT

Fortunately, many Craft teachers have had years of experience in workshop management, so the prospect of organising a room to cater for the demands of GCSE is going to be one of reorganisation rather than entering into something that is entirely new.

Practical subjects are no longer centred on the use of a single material, and yet the large majority of secondary schools in England, Wales and Northern Ireland have been designed to cater for the use of either wood or metal. Even in the 1950s, when open-plan workshops were very much in vogue, all Metalwork was done in one part of the workshop area, Woodwork in another and Technical Drawing in another. It was not long before partition walls appeared and the three specialisations were separated again. Since no change in curriculum thinking had taken place, no wonder there was a move back to what had been successful in the past.

However, considerable curriculum development has occurred since then and many lessons have been learnt about 'open planning'. The present climate of opinion favours the development of 'multi-media workshops'. Besides eliminating the need for major reconstruction of the school building, the idea of one room containing facilities for working in wood, metal and plastics for a single class does away with the difficulties of supervising classes in the much larger open-plan spaces. The workshop should now be capable of allowing a pupil to design more freely and to choose the material or combination of different materials to realise a design under the supervision of one teacher. The alternative situation, which can work quite successfully, is where the workshops retain their specialist identity and the pupils move to the area that suits their particular needs, but this requires block time-tabling and a team of teachers who work successfully together.

'Sketchmate' drawing aid by British Thornton

The Multi-media Workshop

The facilities necessary to do practical work using any one or all three main materials are being brought together in our schools at present. In fact some have gone as far as equipping a room so that all aspects of designing can be completed in it. The drawing boards and stools are as essential as any other equipment and can be used as and when required. Congestion is easily avoided. Pupils inevitably work at different rates and, with the variable demands of each project, it is highly unlikely, even in a full class, that all pupils will want to use a particular piece of equipment at the same time. The exception to this is at the beginning of a year, when pupils are starting a new project. Therefore sufficient boards, stools and drawing equipment should be available for each pupil to use at the same time. With flat boards, i.e. without stands, stacking is not difficult and should present no storage problem. Alternatively pupils could use the British Thornton 'Sketchmate'.

A multi-media workshop should normally be designed to cater for 20 pupils following either CDT Design and Realisation or CDT Technology, so that the pupils are able to work in wood, metal or plastics. The room should have the following facilities as a basis:

2 woodwork benches with 8 working spaces
3 metalwork benches with 12 working spaces
1 brazing hearth
2 pillar drills
1 bandsaw
2 metal turning lathes
1 vacuum-forming unit
2 strip heaters for bending plastics
1 dip coating tank
20 stools
20 drawing boards

It should also have all the hand tools required for working in wood, metal and plastic and the instruments required for drawing.

The Design and Communication Room

Very little modification is needed to make a room that had been used for Technical Drawing into a room suitable for CDT Design and Communication. The major problem is going to be one of finding space to accommodate a table for cutting cardboard, an airbrush bay, and maybe a bench for working with resistant materials. For those with computers and all the associated equipment, the problem of finding space becomes even more difficult, but it is hoped that such facilities can be made available for all students of this strand.

The students are encouraged to use a variety of media and should throughout the course have the opportunity to use: felt-tipped pens, fibre-tipped pens, drafting pens, coloured pencils, and airbrushes if available. Such equipment as flexicurves,

An airbrush being used in a Design and Communication room

French curves, templates for ellipses, circles, flow diagram symbols, adjustable squares, drafting machines, trammel compasses, etc., should be available in sufficient quantitics.

Square grid papers, isometric grids and perspective grids should be available and be part of a standard stock. The move towards having a much wider range of equipment and materials is being made to encourage pupils to extend their graphic skills and to enable them to be selective when responding to a design problem.

The Technology Room

Though in some schools there are rooms containing 'technological' equipment, benches designed to deal with the needs of pupils involved in solving electronics problems, long rows of worktops with various means to supply air pressure, gas, or low-voltage electricity etc., they take on the appearance of a laboratory rather than a workshop. But for those studying CDT Technology as opposed to Technology, there is an important need for wood, metal and plastics resources. Many of the

61

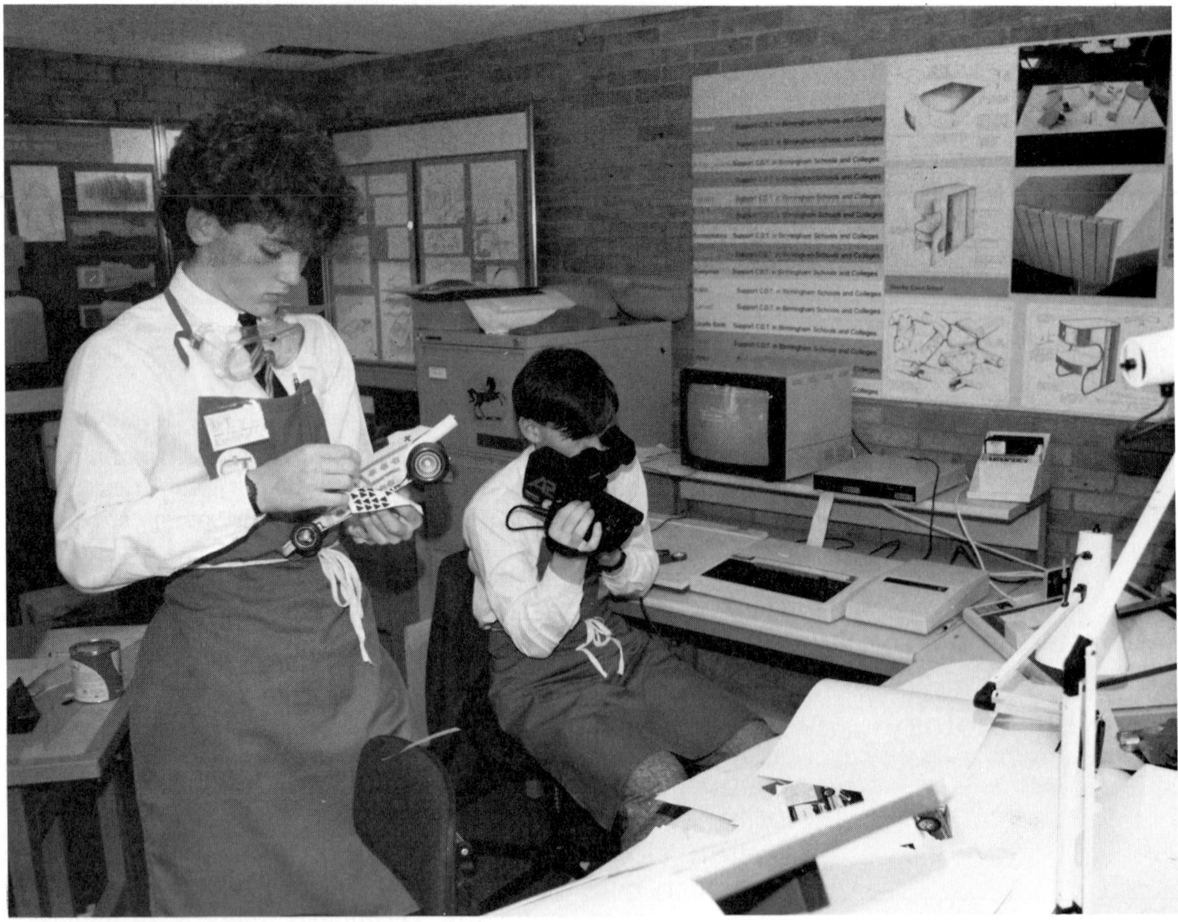

Multi-media workshops

62

projects will need to be resolved using these materials. Therefore the multi-media workshop described earlier would be adequate for much of the practical work needed to complete the common core of problem-solving.

Additional equipment for providing a low-voltage d.c. electricity supply is essential for those pupils studying the electronics module. Also, if not already available, electric soldering irons will be required for many projects, and a bank of between six and ten power outlets could be needed at any one time during a lesson. Construction kits are of considerable value for resolving mechanical problems, and the module on mechanisms could not be attempted if they were not available. The practical application of knowledge is all-important and is a dominant theme in all GCSE subjects, not just CDT.

The Technology Bus

More and more education authorities, supported by industry, are using specially equipped buses to visit those schools which do not have the necessary resources for teaching Technology. These buses are designed to accommodate pupils in very much the same way as any teaching room. In fact they are mobile classrooms with very specialised equipment, and it is only the outer shell that resembles a double-decker bus.

On arrival at a school, at a time when the pupils are due to have their Technology lesson, the bus can be plugged into a mains supply so that the equipment can be operated and the air conditioning put into action.

With each bus there has to be a driver and, more often than

The Isle of Wight Design and Technology bus

not, the driver is a qualified teacher of Technology who has simply obtained an HGV driving licence. So not only does a very specialised mobile room arrive at the school, but also a highly competent and experienced teacher. The advantages of this are tremendous. The pupils' own teacher and the visiting teacher work alongside the pupils, giving them greater individual assistance than could possibly be achieved by a teacher working alone.

At present these Technology buses are in heavy demand and are able to visit a school only once or twice a week. This means of course that facilities within the school are essential for continuity of work during the times when the bus is not visiting. It must be seen as an invaluable support to work already being done in the school and not as an alternative way of educating pupils in CDT.

The Avon Technology bus at Filton High School

4 Planning Schemes of Work

PUTTING AIMS AND OBJECTIVES INTO PRACTICE

Many teachers have faced and still are facing up to the demands of implementing GCSE courses. The task of producing courses and providing activities for 14–16-year-old pupils to enable them to respond to the best of their ability and to 'demonstrate what they know and can do' in coursework is a problem-solving exercise in itself, and something which very few teachers could say that they have been trained to do. The criteria have been established and the specifications laid down nationally, but the development of the teaching programme has become the responsibility of the teacher. Some teachers are working in small groups while others are working individually, hoping to produce something worthwhile. This places a tremendous responsibility upon the teacher, and very few have been confident enough to plan ahead for more than one year. Without the experience and a precedent to follow, much of what is developed can only be looked upon as trial material. Considerable change and development must inevitably take place during these formative years of the GCSE. Until such time as experience can be called upon to influence development, every aspect must be under constant review and, even then, anything which has been deemed to be successful must not be inscribed on tablets of stone because it might undergo further change. In this way the subject will be kept alive and relevant to the needs of the times.

Fortunately for the teacher of CDT, much change in curriculum development has been taking place in recent years that is applicable to many of the proposals and demands of the National Criteria. Problem-solving is a most important educational vehicle in the three strands of CDT and one with which many teachers are familiar. This, then, should be the central theme of all CDT schemes of work.

Developing a Scheme of Work

Problem-solving activity is, as we have seen, central and common to Design and Realisation, Technology and Design and Communication. Therefore a single structure of a model for a scheme of work may be readily acceptable for adaptation and modification to suit the needs of each strand of CDT.

A design process is, more often than not, illustrated and described as a loop, simply because the final stage (evaluating) is carried out by going back to the beginning and referring to the design brief. However, a scheme of work based on a project is more easily illustrated in a linear diagram, with the stages in chronological order.

A project must have a completion date. The period of time required to complete a project is dependent upon the requirements of the syllabus being followed and the stage reached in the course. In general, the projects in the Design and Realisation syllabus tend to take longer than those undertaken in the Technology course, though there are very few precedents on which to make a precise judgement. Generally, the projects in Design and Realisation become more demanding as the course proceeds, in which case more time may be allowed. A review of the syllabuses will show that either a selected sample of coursework is required for moderation or only that work completed during the twelve months immediately prior to the first written examination. Though pupils work at variable speeds, they must all work to the schedule given for the project. The refinements necessary for this to be achieved must be taken into account at the stage of developing ideas. The pupil, in consultation with the teacher, will choose an option which is both a viable solution to the problem and one which can be realised in the time available.

The model illustrated (Figure 1) is planned to be a term's work. Though the length of a term may vary and you may have more time than is allowed for in this model, you should remember that other aspects of the syllabus have to be covered. Also, in the event of an unavoidable absence, there is leeway for catching up with missed work. In addition, other school events may clash with CDT lessons, so the opportunity to have a little flexibility may prove to be most valuable.

The ten-week schedule is sub-divided into parts which correspond with the elements that form part of a problem-solving exercise. The length of time given to each element varies from one week to a period of four weeks. These allocations may vary from one project to another, but the extra time given to one element must be taken from another so that the completion date remains unaltered. This is a valuable experience for pupils to gain while at school and should form an integral part of a project activity. The requirements imposed by each of the three strands in CDT will also influence the proportion of time needed to complete each element. The Technology strand may appear to place a greater importance on research and experiment than on producing an end product, and this will therefore be reflected in the time allocated to each sub-division. The model must be seen as a guide and not as a structure which cannot be changed, modified or refined to suit specific needs.

Many pupils who are being prepared for the 1988 examination have started the course with little or no previous experience in problem-solving. They may have followed the more traditional approach course in CDT or in Technical Drawing and know very little or nothing about tackling problems as such. It is possible, of course, that their motor skills may be more developed than those pupils' with far less experience in making artefacts. Thus for these pupils the first year of the CDT, GCSE course may have to be a time of balancing their skills

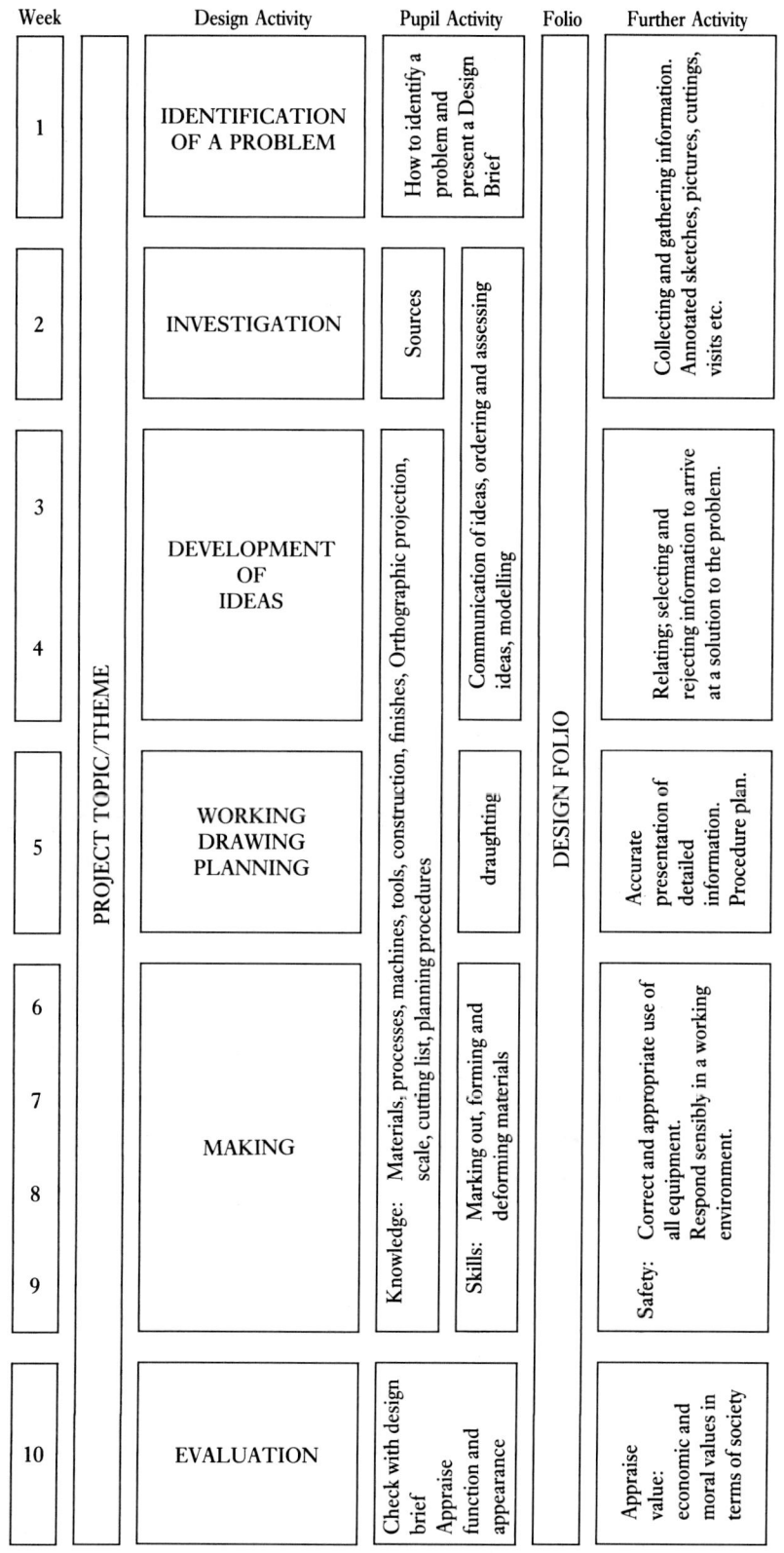

▲ *Fig. 1 Model scheme for Design and Realisation*

profile. In other words, greater emphasis will be placed on design skills and less importance attached to the acquisition of the skills needed for making artefacts. Since not all coursework is required for assessment in many of the syllabuses, an opportunity exists for a 'foundation course' to be built in to the introductory stage. This will then enable the pupils to be adequately prepared for those stages that will produce assessable coursework. The argument could be presented that all pupils should follow an introductory stage in 'designing' to consolidate what has been previously learnt in years 1–3.

Eventually all pupils will have followed a problem-solving course in the first three years and be suitably prepared for proceeding to a GCSE course in CDT. Until such time that we can be confident that curriculum development has reached this stage, the introduction will have to bear in mind the needs of the pupils who may not have been so fortunate.

The first term

It is important that the course gets off to a good start, and it helps from an organisational as well as an educational point of view if the pupils work on a common theme. The first problem-solving activity may be strongly influenced and guided by the teacher. The teacher may also be the initiator of the theme, which should dispel any attempts to suggest other themes and cause delay in getting started. There should be ample opportunity for individual approaches to be developed within the theme.

A Bristol school chose 'A warning device' as its theme for introducing the Design and Realisation strand. The situation where the device could be employed was left to the pupils to decide. They were given three situations by the teacher which could be used for making a choice, but they first had to recommend alternative situations on a printed sheet which they had been given at the beginning of the lesson. The printed sheet also gave guidelines so that one stage was completed at a time and everything was teacher-controlled. The pupils were strongly recommended not to try to think of solutions but to concentrate on situations. While this topic was specifically chosen for Design and Realisation, it would be equally suited for Technology or Design and Communication. Having identified the ten situations, the pupils had to choose the one they were going to attempt. They were advised to think of situations of which they had some experience, whether at home or in school, where warning devices are or could be used.

The choice having been made, it was now essential to express the theme and situation as a design brief. A second printed sheet was given to each pupil to guide him/her through the next stage of writing a design brief. By using an example of a design brief produced to satisfy another situation, the pupils were able to write a brief for themselves and thus were not obliged to copy from an existing one that had been already produced for the situation being attempted.

The next stage of drawing up an analysis had not been reached at the time of writing this text. However, it is plain to see that the teacher's intention is to be very much in control of the activity and to direct every move in the first project. Sound procedures may then be adopted by the pupils to follow in future projects.

The second project

This should be planned to allow the pupil greater opportunity to contribute to the theme and to initiate a design brief. If the pupil can identify a need and produce a suitable design brief, so much the better. For this to be followed through using the structure adopted in the first project the pupil is not only learning and developing design techniques but also acquiring the ability to work more independently. However, even though a pupil may have shown the first signs of independent thinking, he/she cannot be assumed to be capable of continuing this throughout the project. Constant observation of the pupil's progress and guidance on the part of the teacher can never be completely indispensable.

The period of time that can be allocated to the second project will vary according to the requirements of the syllabus. Some syllabuses require the work for assessment that has been done during the twelve months immediately prior to the date of the first examination; others require only a selection taken from the final year of the course. The responsibility rests with the teacher to plan a scheme that is appropriate to the syllabus.

The final project

A pupil can expect to be involved in two to four projects throughout the two-year course. The final project in general should be the project which will best illustrate 'what the pupil knows and can do', so it should be given ample time. The importance of working to a deadline, as mentioned at the beginning of this chapter, must now be fully appreciated, if it had not been done so earlier in the course, for now the final project has to be ready for assessment – not only by the pupil's teacher but also by the external visiting assessor.

The topic or theme for a project may be set by the examining body, in which case any plan or scheme of work must be designed to accommodate such a project. The theme is announced usually at the beginning of the spring term, thus allowing a period of between twelve and fifteen weeks for research to begin and the project to be completed. This aspect is often a compulsory component and will form part of the package to be assessed. Constraints on time such as this must be identified in order to allow other projects to be completed.

SAMPLE SCHEMES OF WORK

The following schemes of work were devised by the CDT department under the leadership of the Head of Department, Kevin Dengate, at the Graham School, Scarborough. As with all schools, the act of committing a scheme of work to paper has not been an easy task, so some aspects have been written in such a way that they allow for a degree of flexibility. This is not to say that the scheme has not been fully developed, but rather that some decisions can only be made when experience has been gained and appropriate stages have been reached. The schemes have been developed in connection with the three strands of Craft, Design and Technology produced by the Northern Examining Association.

Design and Realisation

This course offers an examination assessment in coursework which has a weighting of 50 per cent, and two written papers which together account for the remaining 50 per cent of the marks. It must be remembered that the pupils following this course will have completed a three-year foundation course and be familiar with the design skills that are related to problem-solving, i.e. recognising the existence of a problem, producing a brief, gathering and ordering information, analysing the situation, forming a detailed specification, developing ideas, selecting a solution, realising the chosen idea and evaluating it against the specification.

The pupils are given a copy of the Coursework Assessment Sheet (Figure 2) so that they have not only a reminder of the essential features of an activity in which they will be involved when solving a problem but also a guide to enable them to plan their work.

The presentation of their research material and design thinking is left to the discretion of the pupils. This allows a personal approach to be developed, so that the final product of the folio will be a reflection of a pupil's character, innovation and motivation. The pupils in the lower ability range may follow the Coursework Assessment Sheet more rigidly than more able students and find it a helpful structure to guide them and give them confidence to tackle a problem-solving experience.

Communication skills

Though these skills have formed a central element of the foundation course in years 1–3, it will be necessary to reinforce these basic skills with further work and experience in three-dimensional drawing, rendering techniques and the production of working drawings. Rather than impose a set of exercises to develop these skills, all the work done in this connection will be covered during the pursuit of solving a problem and will be manifested in the design folio.

Knowledge

As with communication skills, the pursuit of knowledge will be extended as a result of the problem-solving experience. However, not all aspects of knowledge can be expected to be covered through project work alone, and thus it will be necessary to

Candidate's Name.. No.		M = MODERATOR'S MARK T = TEACHER'S MARK	MARK	PROJECT MARK		Candidate's Name..............

SKILLS	LEVEL OF RESPONSE	MARK RANGE	M	T
DESIGN **1 Recognition of Problems** Identification and formulation of a brief worthy of investigation	No suitable area of study identified	0		
	Considerable guidance needed	1-2		
	Some help or guidance needed	3-4		
	Problem identified unaided	5-6		
2 Research A valid self directed and ordered investigation	No research presented	0		
	Narrow range of research gathered with little purpose	1-4		
	Mainly relevant information gathered but some assistance and/or redirection needed	5-8		
	Broad and imaginative research planned with little or no assistance	9-12		
3 Analysis Recognition of relevant social, physical, functional material, aesthetic & economic factors	No analysis presented of the problem or the research material	0		
	All relevant factors not considered *or* considered to a shallow level	1-4		
	All major factors considered to some extent but not all to sufficient depth	5-8		
	Logical presentation of problem; all relevant factors considered in depth	9-12		
4 Specification Interpretation of problem in a clear definitive statement which gives details of constraints	None drawn up	0		
	Specification misdirected, considerable guidance needed	1-2		
	Generally correct but not too clear and/or some assistance needed	3-4		
	Concise, definitive, drawn up with reference to research and analysis	5-6		
5 Generation of Ideas The unaided production of a variety of tentative ideas	No ideas or alternatives to that used in the solution	0		
	Little variety or only variations on one proposed idea or solutions that do not meet the criteria of the specification	1-6		
	Limited variety that meets most of the criteria of the specification	7-12		
	A wide range of alternative proposals with distinct differences that meet all the criteria of the specification	13-18		
6 Development The selection and development of one idea to a detailed workable solution	No recognisable solution	0		
	Few reasons for selections of idea; little further development of design detail, poor communication with confused or few details	1-8		
	Some reasons for selection; limited development of design detail and range of communication; main details of size, materials, construction given	9-16		
	Valid reasons for selection, thorough development of design details good range and standard of communications with clear details of size materials and construction.	17-24		
7 Evaluation The testing of the solutions and estimation of how well it satisfies the problem	None attempted	0		
	Irrelevant or largely superficial	1-4		
	An honest attempt which lacks objectivity, either incomplete or, in parts, irrelevant	5-8		
	Thorough, relevant, concise, objective	9-12		
	(A) DESIGN SUB-TOTAL(S)	90		
8 MANUFACTURING **(a) Construction** Workmanship. A range of appropriate techniques	Not submitted (if no realisation is submitted no grade will be determined)	ABS		
	Simple or inappropriate construction showing few techniques and/or a poor standard of workmanship	1-10		
	EITHER A limited range of simple or appropriate techniques, but showing a good standard of workmanship *OR* a larger range of more complex techniques showing a poor standard of workmanship	11-20		
	A wide range of more complex, sound, constructional techniques showing a high level of workmanship	21-30		
(b) Accuracy In relation to working drawing and constructional details	Little accuracy	0-2		
	Overall dimensions generally accurate, some details not as planned	3-6		
	A high degree of accuracy with all details as planned	7-10		
(c) Finish	No surface finish or inappropriate finishes	0-2		
	Appropriate surface finishes, but details of limited quality	3-6		
	High standard of appropriate finish to all materials	7-10		
(d) Aesthetic Quality Judgement of appearance function and suitability for its environment	Poor appearance, unsuitable for function or its environment	0-2		
	Generally suitable but limited in some details	3-6		
	Appearance and function highly suitable for its environment	7-10		
(e) Applied Knowledge Transfer of design proposal into planned sequence of actions. Correct use of appropriate tools, techniques and processes	Supervision of all work needed	0-2		
	Understands general use of tool techniques and processes but some guidance needed	3-6		
	Most realisation processes planned and executed with little guidance	7-10		
	(B) MANUFACTURING SUB-TOTAL	70		
	(C) ADJUSTED DESIGN SUB-TOTAL (Divided by 3)	30		
	(D) PROJECT TOTAL	100		
			M	T

▲ *Fig. 2 Design and Realisation Coursework assessment sheet, NEA* augment the learning process by using worksheets and setting homework assignments. The department has a valuable collection of worksheets relating to the areas of knowledge concerning the manipulation, characteristics and properties of materials which are readily available for use when needed. Other reference material related to energy, control, movement, adjustment, holding, locking, joining, forces, structures, wear, surface treatment, ergonomics, aesthetics, and design in society, is also available as resource material which, coupled with the information sheets from the Technology and the Design and Communication strands, forms a complete range of learning material for all aspects of the course.

The worksheets form the basis of the fourth and fifth year homework assignments. Each of the topics listed will be introduced more formally at intervals throughout the course, using teaching sessions which will involve demonstrations, the use of film and video. Knowledge will be assessed periodically by tests that require responses to short-answer questions. The order in which each topic is taught or tested is not predetermined and will arise as a result of consultation with each member of staff in the department.

Coursework

At the end of the third year, when all the pupils have made their course choices in GCSE and know which strand of CDT they are taking, they will be set tasks which will involve them in identifying a problem, outlining an analysis and gathering relevant information. The responses to these tasks will then form the basis of the introductory work that begins in the first term of the course in the fourth year. The pupils will work on a common theme and be guided through the appropriate design approach in order to arrive at a viable solution. Considerable emphasis will be placed on the need for developing communication skills and completing the designing activity in the first term. Observation skills will be developed through homework assignments that require a close examination of common household appliances.

The second term is envisaged as a time when the realisation and evaluation of the introductory project is completed. It is essential that during this time very close attention is given to the need for the pupils to have individual tutorials in order to assess accurately what has been learnt and what has been shown. The information gathered in such an exercise will also be invaluable for giving guidance on further development and for identifying particular strengths, interests etc., which may influence the choice of the major project. The introductory project must be completed by the end of the second term to allow the twelve months required to complete the assessable coursework project or projects.

During the period leading up to the end of the second term, time will be given to individual tutorials to guide pupils with their choice of topic for a major project. Each pupil will be given the opportunity to identify a problem that will be sufficiently demanding and appropriate for the course being followed. The use of diaries or project planners will be used to encourage good management of time so that deadlines are met and the project work is completed within the twelve month period immediately prior to the first written examination. It is hoped that the more able pupils using a diary or planner will be able to plan their own strategies for designing and making, and will meet their own self-imposed deadlines.

Design and Communication

This course offers an examination assessment in coursework which has a weighting of 40 per cent, a design assignment that has a weighting of 30 per cent and a written paper that also has a weighting of 30 per cent.

The minimum requirements for any pupil following this course to be awarded a coursework grade is that they should demonstrate their Design and Communication skills in each of two topics and establish a clear relationship between the original problem and the solution. They should also complete a design assignment and a written paper. Omissions in any section will result in the pupil being recorded as absent.

Skills

This section is divided into three areas: design skills, making skills and communicating skills.

The design skills needed to fulfil the demands of this syllabus are identical with those required to fulfil the demands of the Design and Realisation and the Technology syllabuses. The pupils are required to: identify a problem, gather and order information, analyse a situation and foresee the implications of different approaches, draw up a detailed specification, develop ideas, synthesise the results into a feasible proposition, arrive at a viable solution to the problem, and evaluate the outcome at the drawing-board stage of manufacture. Therefore the structure of this course will be almost identical to the one outlined for design and realisation.

The pupils are given a copy of the Coursework Assessment Sheet (Chapter 5, Figure 5) as an outline of what is involved and for them to use as a guide when tackling a problem. The Design and Communication course requires a slight change of emphasis in that the pupils also have to be aware of:

the relationship between designer, consumer and society, and the importance of clear, effective communication methods between them, the use of graphic communication methods in the design process as a means of developing ideas and proposals, the potential of three-dimensional forms as a means of communicating information and ideas, and solutions to problems and sub-problems.

The syllabus states that 'there is a continuous gradation of graphic skill, from formal Orthographic projection at one extreme to impressionistic illustration at the other'. The CDT Department feels that it is essential that these four elements are treated together rather than as separate topics. Throughout the foundation course pupils will have experienced aspects of all four elements, albeit at an elementary level. The examination course must aim at reinforcing this basic work. It must also provide an opportunity for the four elements to be experienced in projects so that communication skill are related as a whole. Thus, a project concerned with the design of a chess set can be an assignment to develop aspects of pictorial drawing while at the same time provide an opportunity to develop aspects of Orthographic projection and impressionistic illustration.

A project primarily concerned with illustration, e.g. the analysis of the construction of a pump-action toothpaste dispenser, will inevitably combine aspects of pictorial and Orthographic representation. Pupils must acquire the skill to recognise the relationships of these skills and be selective when choosing an appropriate method of illustration. Central to this activity is the notion of designing. Therefore all projects must be suitably chosen to encourage a problem-solving approach and to help pupils reach a stage where they can identify the two suitable coursework projects.

The syllabus states that 'candidates will be expected to have experience of making, modelling and using mock-ups as a means of exploring a problem and as a means of testing the feasibility of a solution'. Since these experiences are an integral part of the course, all pupils will be expected to be involved in making a scale model or a mock-up to illustrate an idea or a solution. These will arise as a result of a problem-solving activity rather than as a formal exercise in making models.

Knowledge

The knowledge content must be seen as that which is essential to developing both Design and Communication skills. The teaching of specific topics will be necessary, but the precise order will not be decided until the nature of the projects has been determined. The aspects of knowledge listed, i.e. materials and components, mechanical control and energy, will be largely dealt with by the use of resource sheets, either of a specific Design and Communication nature or drawn from the other CDT strands. Where possible, these elements will be linked to the other more familiar areas of knowledge, construction methods and applied geometry to the development of projects. The aim will be to make the pupils aware of the broad spectrum of relevant and related knowledge.

Coursework

The timing and organisation of the two coursework projects will follow a similar pattern to that for Design and Realisation. However, since the spring term of the final year of the course is allocated to the design assignment, the two individual course-work projects must be completed by the end of the Christmas term of the second year. The coursework projects, while being of an individual nature, will contain common themes relating to the knowledge content of the course. These themes will need to be identified and used as vehicles for the teaching of specific skills not previously covered by the introductory projects. As with the Design and Realisation course, the use of project diaries or planners is to be encouraged so that the pupils can determine their own strategies to achieve deadlines for completion of projects.

Technology

This course offers an examination assessment in coursework which has a weighting of 50%, and two written papers that have an equal rating of 25% each.

The minimum requirements for any pupil following this in coursework is that they should show evidence of development work as recorded in a project report and its subsequent realisation. If the pupil fails to attempt an element, an 'absent' mark will be recorded.

Skills

As with the CDT Design and Realisation and the Design and Communication courses the pupils will be required to demonstrate their abilities in: recognising a problem, analysing and identifying the factors which influence ergonomics, functional modes, environmental considerations, gathering and ordering information which will stimulate ideas, writing a detailed specification outlining constraints imposed by knowledge, resources, cost etc., developing a variety of ideas and proposals for a solution, selecting a solution on the basis of reasoned arguments, presenting constructional details in sufficient depth for a competent third party to construct, and evaluating the final solution or proposal against the specification. The Coursework Assessment Sheet will be used as the guideline for the pupils (Figure 3).

It must be remembered that pupils following this course will have completed a three-year foundation course and will be familiar with the skills that are related to problem-solving activities. Thus they will be familiar with the terms used above. Similarly, it is expected that the pupils will have reached a basic level of competence in the manipulation of hand tools, materials and components commensurate with achieving a satisfactory standard of accuracy, finish and function in the realisation of technological projects. Communication skills, especially written and graphical, will be emphasised, and a short programme of exercises will be undertaken. These exercises are designed to reinforce the skills of presentation, use of colour, diagrams, and the use of signs and symbols covered in the foundation course. It

Candidate's Name... No.					M = MODERATOR'S MARK T = TEACHER'S MARK		MARK RANGE	Project Mark		Candidate's Name............................... No.

Schedule	Levels of Attainment	MARK RANGE	M	T
1 Recognition of Problem The formation of a brief worthy of investigation.	No suitable area of study identified.	0		
	Superficial indication of what the project is to achieve.	1		
	Good presentation with help given.	2		
	Unaided detailed indication of what the project is to achieve.	3		
2 Analysis of Problem The identification of the factors which influence the problem (e.g. physical, social, functional, economic, environmental).	No analysis.	0		
	A very superficial analysis.	1		
	Individual sub-systems identified.	2		
	The individual sub-systems identified and *some* considered in detail.	3-5		
	The individual sub-systems identified and all considered in detail.	6-7		
3 Research/Testing The gathering and ordering of information related to the brief.	No relevant research.	0		
	Some relevant research from only one source.	1		
	Some relevant research from several sources.	2-3		
	Research relevant to most of the analysis from several sources.	4-5		
	Research relevant to all the analysis from a range of sources.	6		
4 Specification A detailed description of the requirements of the project.	No specification presented.	0		
	Only broad areas indicated.	1		
	Precise areas indicated.	2		
	Precise definition of project given with relevant parameters.	3		
5 Generation of Ideas The stage at which candidates give expression to proposals which meet the problem or its parts.	Only one solution considered.	0		
	A variety of solutions considered.	1-6		
	Sub-systems identified in outline.	0-2		
	Some sub-systems analysed in some detail.	3-4		
	Some sub-systems analysed in detail.	5-6		
	All sub-systems analysed in detail.	7-8		
6 Solution *(a) Level of solution*	A haphazard treatment of a low level problem.	0-2		
	A logical treatment of a low level problem.	3-4		
	A superficial treatment of a medium level problem.	5-6		
	A superficial treatment of a high level problem.	7-8		
	A logical treatment of a medium level problem.	9-10		
	A logical treatment of a high level problem.	11-13		
(b) Constructional Skills	Nothing of value has been made.	0		
	An incomplete assembly of one or more sub-systems. Constructional skills minimal and/or inappropriate.	1-5		
	The assembly of several sub-systems is complete, but choice of materials poor and/or assembly of components inadequate which precludes assembly from working correctly.	6-10		
	The solution is constructed so that it meets the specification. It will perform the intended use and is neatly assembled.	11-18		
(c) Materials	Sensible choice of materials available in centre.	0-2		
	Sensible choice of materials from beyond centre's resources.	3-5		
(d) Safety	Some aspects of safe use of materials and equipment neglected. Sensible approach to own safety.	0-2		
	Sensible approach to own and others safety and materials and equipment used.	3-4		
7 Presentation of Report *(a) written*	No written report presented.	0		
	Some written evidence presented.	1		
	Relevant well organised report.	2-4		
(b) numeric	No evidence of quantitative testing.	0		
	Some evidence of quantitative testing.	1		
	Use of appropriate data and calculations.	2-3		
(c) graphical	No graphical work presented.	0		
	Some aspects of the report well-illustrated.	1-3		
	The entire report is well-illustrated.	4-7		
(d) modifications	No modifications recorded.	0		
	Testing of sub-systems for validity.	1-2		
	Outline of tests recorded.	3-4		
	Record of modifications made as a result of tests.	5-6		
8 Evaluations *(a)* **by** *the candidate*	No evaluation is presented.	0		
	A detailed evaluation against part of the specification.	1-2		
	A detailed evaluation against the complete specification.	3-4		
(b) **of** *the candidate*	The candidate: could not proceed unaided;	0		
	made most of the decisions;	1-2		
	proceeded unaided throughout.	3		
	PROJECT TOTAL	100		
			M	**T**

▲ *Fig. 3 Technology Coursework assessment sheet, NEA*

is envisaged that in this and aspects of material choice and manipulation the Technology course will make use of appropriate resource material from the Design and Realisation and the Design and Communication courses.

Knowledge

The syllabus provides a coherent and detailed outline of the knowledge content of the course as defined by a systems approach. It is the intention that during the fourth year pupils will cover aspects of Technology relating to the following areas:

a) Electronics;
b) Structures and mechanisms;
c) Timers and computer control;
d) Logic;
e) Pneumatics.

Each element of the course will be covered by a problem-solving approach that is based on the notion of globules of knowledge, e.g. linking work done in the Science Department with that covered by the CDT Department in basic electronics and switching. Additional information and grounding in the use and manipulation of basic electronic components (transistors, resistors, diodes, relays, capacitors, etc.) and circuitry will be provided in the form of factsheets, worksheets and other support resources. The basic principle of a systems approach (input – process – response) is to be covered, together with the skills of collecting and recording information and the construction of circuit diagrams.

Simple problems will be set requiring the manipulation of various facts, components and materials.

The teaching group will be divided into two sub-groups, and problems related to the topic under consideration will be set to smaller working groups within each sub-group. The problems will be defined so that gaps of knowledge will be highlighted, and the working groups will be expected to fill these gaps by using the appropriate resources; in so doing they are expected to increase their knowledge and understanding. It is essential that the information acquired and the skills developed are recorded during this process so that they can be used to solve problems of a more complex nature.

The continual setting of problems that match the increasing levels of skill and knowledge demonstrated by pupils will create new knowledge gaps that require filling if progress is to be made. The process is spiral in that, as the pupils' ability improves, so the range of knowledge and learning to manipulate materials and tools is extended. It is envisaged that in this way pupils will generate their own bank of skills, interests, knowledge and resources that will enable them to formulate and solve practical technological design problems for their major course-work project to be completed during the final year of the course.

Coursework

The approach to the formulation of viable coursework projects will be similar to that outlined for the Design and Realisation and the Design and Communication courses. The need for careful guidance at each new stage and the need for individual counselling cannot be overemphasised, so it is intended that tutorials and guidance will form an essential ingredient of this course. The process of counselling will begin approximately halfway through the summer term of the first year of the course. Pupils will be expected to have isolated the area of the problem they wish to pursue by the end of the term so that they can begin the process of analysing, writing a detailed specification and gathering relevant information during the holiday period. In common with the other CDT courses the use of project diaries or planner is seen as an essential aid to the completion of the coursework project within the deadlines set by the examining body.

Summary

Central to each of the schemes described in this chapter is the activity of solving a problem. The areas of knowledge and skill are also recognised as essential ingredients but, where possible, they are used as support activities for each project that is attempted. The further the pupil moves along each stage of the course, the more demanding each project becomes and the areas of knowledge and skill more extended. Much of the research needed to make a project viable takes a considerable time to assemble, so in order that deadlines may be achieved pupils must carry out this type of work in their own time; this means taking advantage of holiday periods as well as homework assignments.

Pupils must become familiar with using a diary or a planner, since they need to perceive an assignment as an activity that is to take place over a relatively long time, and that a problem-solving activity is made up of many much smaller assignments all of which are totally dependent upon each other.

The teaching or learning resources will need to be readily available in two main forms: as information packs and as worksheets. Since individual pupils will require resources which relate to their projects at varying times, it is essential that a filing system is adopted to cope with these needs.

Time must be allowed for group and individual tuition. It must be remembered that the teacher's role has considerably changed and that today's teacher is committed to tutoring, monitoring and assessing as well as guiding pupils through an externally assessed course in GCSE.

5 ASSESSMENT

ASSESSMENT OF COURSEWORK

Some indication of the coursework requirements were given in Chapter 1, but now it is necessary to concentrate on the task of assessment. This must be done continuously throughout all or part of the course. A good time for assessing what has been achieved usually comes at the end of a project, but this is easier said than done. However, it is good practice to review what has taken place as near to the time of completion as possible so that weaknesses and strengths can be identified and future remedial methods can be planned. It is important that the student is involved in the assessment procedure so that there is no doubt about expectations and levels of performance. In this way the student has no illusions about progress and has the opportunity to direct effort to those areas where improvement could be made.

Ideally the assessment procedure should be conducted by the teacher and the student in the absence of other distractions and with the complete package of coursework available. The occasion should take on a friendly atmosphere, be seen as natural continuation of learning and certainly not one of examiner and examinee in an examination room. The process from the inception of a project to the final assessment should be seen as a normal and complete course of events. In this way the assessing time, it is hoped, will not be an ordeal for the students but just another integral part of a learning process.

KEEPING A RECORD

A record of events and responses must be made for every student. Each teacher will need to devise a system so that a record can be made of the students' responses to the assessable objectives outlined by the Department of Education and Science in their document *National Criteria – Craft, Design and Technology.*

Figure 1 shows an example of a 'record of achievement' sheet. In its present detailed form it could be very off-putting for the student to follow and understand. However, it should prove helpful to the teacher, and it can form a guide for developing a more simplified record sheet similar to the example illustrated in Figure 2 for the student to use.

Using the Detailed Record of Achievement Sheet

Each aspect of the design activity is listed in the second column. The figures in the first column refer to the Assessment Objectives. Some of these appear more than once and are by no means fixed. Depending on the project, the Assessment Objectives could be arranged differently or even left out, e.g. 17,

ASSESSMENT OBJECTIVE	DESIGN ACTIVITY	1
3, 4.	**IDENTIFICATION OF A NEED** Recognition that a problem exists Clear statement expressed as a problem to be solved (Design Brief) Analysis-precise details of the problem (specification)	Able to follow a given design brief but requires some assistance.
5, 7, 17.	**RESEARCH** Observe & record detail (Annotated sketches) Primary research (Visits & examination by first hand experience) Secondary research (Reading & Collecting pictures from magazines Empirical research (Learning by experience with materials).	Relies very much on secondary sources only.
5, 6, 8, 9, 11.	**DEVELOPMENT OF IDEAS** Drawing Modelling Experimenting Selection & rejection of ideas on the basis of feasibility & relevance to the design brief. Recognition of constraints – limitation of materials manufacturing resources and personal manipulative skills. Also limitation of time and cost.	Has a single idea & requires assistance with choosing materials & choice of appropriate construction techniques. Suggests minor modifications to proposed design for economic reasons. e.g. time & cost.
2.	**COMMUNICATION OF IDEAS** Convey information graphically and sufficiently accurately to make production easily understood.	Producers images on isometric grid paper to give elementary detail.
12, 16	**PLANNING** Arrange a sequence of operations to permit efficient production of the solution. allow for possible refinements and the constant need to refer to the design brief.	Works at an adequate speed and hopes rather than plans to finish on time.
13, 14, 16.	**MAKING** Working accurately to the detail and sizes given in the working drawing with due regard for appropriate techniques and safety.	Able to apply basic manipulative skills to realise chosen design.
10, 14, 15, 17.	**EVALUATION** Make critical judgements about the end product in terms of its function, appearance and make recommendations, where appropriate for refinements and or modifications.	Aware of some aspects related to function and appearance.

STUDENT   CENTRE   PROJECT TITLE

CDT COURSEWORK

▲ *Fig. 1 CDT Coursework record of achievement*

'describe the interrelationship between design/technology and the needs of society' does not form an integral part of every project. But a project on conservation of energy – wildlife, materials etc. – would lend itself very much to research and evaluation. The grade criteria in the remaining four columns do go some way towards describing the expectations at four different levels. Number 1 level describes the possible expectations of a grade F candidate. Number 2 level describes the possible expectations of a grade D and E candidate, grade C is described in column 3 and column 4 covers the A–B range. An attempt to produce seven descriptions at this stage would be a little premature, but no doubt when the GCSE has been

2	3	4	
Shows some understanding of a given problem and able to produce a simple specification.	Recognises a need and able to draw up a design brief and specification.	Appraises a situation with understanding and produces a reliably detailed design brief and specification.	
Gathers & records secondary information. Makes basic observations about available & relevant proven solutions to similar problems.	Selective about the information gained through primary & secondary sources. Well organised and documented detail relevant to the problem.	Recognises the broad implications of the problem & systematically probes for the relevant information to satisfy ergonomic, aesthetic, functional & social needs.	
Tries more than a single idea but usually based on a single theme. Considers feasibility at an elementary level & makes some recommendations for improving design proposals.	Examines alternative ideas and concepts but generally relies on well proven solutions which could be adopted, with minor modification, to fulfil the requirements of the design brief. Shows an awareness of cost time and efficiency.	Prepared to develop new concepts through modelling & empirical research. Occasionally innovates a different approach to a problem. Reveals inventive or imaginative thinking & a flair for solving problems. Recognises the potentiality of an idea in respect of fulfilling a design brief.	
Produces 3D images on plain paper & at least two views in orthographic projection to scale.	Uses two or three techniques to give details relevant to form and construction.	Uses a wide range of techniques and provides all the relevant detail for production.	
Works at a reasonable rate with evidence of some planning to avoid delays	Fairly well organised and most aspects considered.	Able to keep the deadlines due to careful planning and an ability to foresee potential snags.	
Applies a limited range of manipulative skills to realise chosen design.	Applies a moderate range of manipulative skills confidently and accurately with due care for safety.	Reveals a degree of accuracy and a 'feel' for the materials used in the manufacture of the chosen design.	
Makes statements, with reasons, about aesthetic and functional qualities of the end product.	Makes valid judgements showing an awareness of the finer points concerning functions, form & appearance. Makes some recommendations for improvements.	Fully aware of functional and aesthetic qualities and makes sound recommendations for improvements & future developments.	

RECORD OF ACHIEVEMENT

functioning for a few years the descriptions will be refined in the light of experience and the seven levels will be more expertly identified.

At present there is no reference to marks. The examination groups, though covering the same design loop, give slightly different weightings, and it will be necessary to relate these to the strand and the particular examination group being used. However, in the early stages the record of achievement can be expressed as levels of performance. For example, a student may show some understanding of a given problem and be able to produce a simple specification, so the box to the right of this description can either be shaded or ticked. For research the

STUDENT						CENTRE					DATE
DESIGN TOPIC:	LEVEL					TEACHER INVOLVEMENT				COMMENTS	
	NO WORK	BELOW AVE.	AVERAGE	ABOVE AVE.	EXCELLENT	CONSIDERABLE	MODERATE	HARDLY ANY	NONE		
	0	1	2	3	4	1	2	3	4		
IDENTIFICATION: BRIEF											
RESEARCH											
DEVELOPMENT: IDEAS											
COMMUNICATION											
PLANNING											
MAKING											
EVALUATION											
SUB-TOTALS										TOTAL	

▲ *Fig. 2 A simplified version of a record of achievement*

student's performance may be most accurately described in column 3, so that the box to the right of this description may also be ticked or shaded. With each design activity receiving an assessment description it should be very easy for teacher and student to see a pattern that shows strengths and weaknesses. For each project there will be a record of achievement sheet, so that a comparison can be made to identify progress or even a failure to maintain earlier achievements.

Some Regional Examination Groups only require a record of the work done in the twelve months prior to the examination date, which means that this type of assessment need not really get under way until March or April of the year prior to the examination.

Using the Simplified Record of Achievement Sheet

This sheet has two main advantages. First, the level of performance can be translated into a mark and the mark for each design activity fulfilled contributes to an overall mark, and, second, the teacher's involvement can be recorded. Thus the pupil who is able to work conscientiously and use the teacher as a consultant can score more highly than the pupil of equal ability who depends largely upon the teacher to be the prime mover. This record of achievement can be filed by the pupil as well as by the teacher, and if used wisely can act as an incentive for improvement as well as guidance to show where improvements could be made. This level of responsibility is to be encouraged in every pupil.

Personal Profile

There are other features that have not yet been considered that are important and will help to tell us more about the pupil. These are more readily identified in group projects where a working relationship with peers is essential in determining the success or failure of a project, but they are also integral to the success or failure of an individual project where a sharing of resources is unavoidable. It is intended that this type of assessment be used as and when the teacher believes it will be helpful to the pupil. It could take the place of an end-of-term or year report or serve as a useful complement to an existing written report. The main purpose, however, is to ensure that the pupil is aware of the complexities of a situation in which learning takes place, especially when involved with project work.

Again the layout presented in Figure 3 is only a sample and should be modified to suit particular needs. Furthermore it is not intended to suggest that this type of recording is required by each of the examining bodies; it is only intended to be of assistance to those teachers who have not yet devised a system that suits their individual needs. The section that follows will show a number of different methods of recording information that is most definitely required to ensure uniformity for cross-referencing purposes.

STUDENT		CENTRE					DATE				
PERSONAL QUALITIES	SELF ASSESSMENT					TEACHER ASSESSMENT					AGREED ASSESSMENT
To be completed in three separate stages 1. By the student. 2. By the teacher. 3. By student and teacher together to agree on a final assessment.	VERY WEAK	BELOW AVERAGE	AVERAGE	ABOVE AVERAGE	EXCELLENT	VERY WEAK	BELOW AVERAGE	AVERAGE	ABOVE AVERAGE	EXCELLENT	
	1	2	3	4	5	1	2	3	4	5	
COOPERATIVENESS Willing to work with others											
RELIABILITY Can be trusted to use time sensibly											
RESPONSIBILITY Prepared to be accountable											
INGENUITY Inventive, imaginative, creative											
INNOVATIVENESS Ability to introduce new ideas											
SUB TOTALS											
					AGREED TOTAL						

▲ *Fig. 3 A personal profile sheet*

Coursework Assessment Forms issued by Examination Groups

Central to all project work is problem-solving, but there are slight variations in the way each strand of CDT has been developed, and this is naturally reflected in the design and format of assessment forms. Some formats have been cautiously and rightly labelled 'specimen forms'. This is to ensure that what has been published is still liable to further development and even complete revision.

The form in Figure 4 can only be completed by following a detailed scheme in which marks are awarded for low achievement, medium achievement and high achievement. Each of these three levels is then further sub-divided to establish whether the performance is best described by the top range or bottom range. For example, in a medium level the bottom range of marks allocated for specification and analysis of a problem is 11–20, while in the top range the marks are 21–30. This method of translating descriptions into marks is fairly standard and most helpful in establishing a reliable assessment of performance.

Because there is sometimes a need to compare the marks awarded by one teacher with another teacher in the same school (internal moderation), some forms are designed to allow an initial mark to be recorded and a revised mark to be added later,

SPECIMEN FORM

MIDLAND EXAMINING GROUP
Joint GCE 'O' Level/CSE Examinations

CDT : TECHNOLOGY 1987
ASSESSMENT OF MINI-PROJECTS (SCHEME A)

Centre Name ....

Centre Number ....

1 The two mini-projects submitted by each candidate must be marked in accordance with the marking scheme in the syllabus.

2 The mini-projects must be completed and marked by 1st May 1987.

Candidate's Number (last 3 digits only)	Candidate's Name (initials and surname)	PROJECT	DESIGN SKILLS			MAKING SKILLS	TOTAL MARK FOR EACH PROJECT	TOTAL FOR EACH CANDIDATE
			1	2	3	4		
		Brief Description	5%	10%	10%	25%	50%	100%
1		1						
		2						
2		1						
		2						
3		1						
		2						
4		1						
		2						
5		1						
		2						

▲ *Fig. 4 A coursework assessment form (MEG)*

should this be necessary. Where there is only one teacher involved in the assessment, it may still be necessary to review the initial marks in the light of the experience that had been gained since they were first awarded.

The design of the form in Figure 5 also allows the visiting Moderator to agree with the marks awarded or make what adjustment is felt necessary to make the marking agree with the National Standard. If there is an adjustment, the teacher should not feel upset that the marking does not quite agree with the National Standard; provided the teacher has been consistent in awarding marks, the amendment procedure adopted by the Moderator is quite straightforward. Inconsistent marking, on the other hand, could lead to a complete re-mark of all the candidates' work. This would be extremely time-consuming and could prove to be an expensive exercise.

Once the moderation has been completed by the visiting Moderator, it may or may not be possible, depending upon decisions still to be made, that the teacher and Moderator consult together about the final marks. This has been the practice with many GCE and CSE CDT-equivalent coursework assessment procedures for many years, and it is one that has proved to be most valuable in establishing a fair assessment of the candidate's responses.

ASSESSMENT BY EXAMINATION

Much time has been devoted to coursework because it carries the heaviest weighting of marks, and also because the teacher is very much involved with initiating, planning and assessing projects.

The other modes of examining only involve the teacher with the preparation of pupils; the actual marking or assessing is carried out mainly by a team of examiners who are quite detached from the school and the candidates being examined. Most examiners are practising teachers who have had a minimum of five years' teaching experience and have demonstrated their competence to mark scripts efficiently and accurately.

Most examination groups offer a similar pattern of assessment in terms of weighting and number of papers. Paper 1 is usually presented in the form of short and long questions to be answered by candidates under examination conditions, and Paper 2 is presented in the form of a design problem that has to be resolved in the space of approximately 2 hours, again under examination conditions.

Paper 1: Technology

This is sometimes referred to as a written examination that has to be taken in a strictly controlled set of circumstances where reference material is unobtainable and communication with other candidates is not permitted – a situation which is all too

Candidate's Name.. No.

I = INITIAL MARK
R = REVISED MARK (IF ANY)
M = MODERATOR'S MARK

Schedule	Level of Response	Mark Range	Topic 1 M	Teacher I	Teacher R	Topic 2 M	Teacher I	Teacher R
1 Recognition of Problems *The candidate has:* Has the candidate demonstrated the ability to identify a problem and/or formulate a brief worthy of investigation? Was this done unaided?	not identified a suitable area of study;	0						
	had considerable help;	1						
	had no help but the resulting problem is unsuitable for further development;	2						
	identified and described unaided a problem but failed to develop it;	3-4						
	identified and described unaided a problem and has then proceeded to develop it.	5-6						
2 Research *The candidate:* To what extent has the research been (a) valid (b) self-directed (c) ordered?	has not presented any research;	0						
	has presented copied material only, without any evident purpose;	1-3						
	did not independently initiate or maintain a planned programme of research but required considerable assistance and frequent re-direction;	4-6						
	required significant assistance initially, but thereafter little direction, concentrating on a narrow area or covering a wide field less relevantly;	7-9						
	has produced relevant and largely self-directed research covering a wide area;	10-12						
	planned a broad and imaginative research programme entirely without assistance and developed it progressively, the resulting information being structured and effectively communicated.	13-15						
3 Analysis *The candidate has:* To what extent has the candidate recognised relevant factors and foreseen the implications of different approaches?	presented no analysis either of the problem or of the research material;	0						
	limited his or her analysis to choosing one solution from previously gathered information;	1						
	limited his or her analysis to choosing more than one solution from previously gathered information;	2						
	made some attempt to break the research data down into manageable units at a superficial level;	3-5						
	produced a logical breakdown of his or her research;	6-8						
	planned at least one approach to producing a solution;	9-10						
	planned more than one approach to producing a solution.	11-12						
4 Specification *The candidate's specification:* Has the candidate attempted to interpret his or her brief in the form of definitive proposals having regard to such considerations as materials, cost, time and the needs of society?	has not been drawn up;	0						
	is unspecific;	1						
	is too wide;	2						
	is definitive and is related to the brief;	3-4						
	is definitive and related to the brief but has been drawn up with obvious reference to the preceding research and analysis.	5-6						
5 Development 5.1 Number of ideas *The candidate has produced:*	no ideas or no alternatives to the one used in the solution;	0						
	one alternative idea;	1						
	two alternatives;	2						
	more than two alternatives;	3						
5.2 Variety of ideas. *The candidate has:*	displayed no variety of ideas;	0						
	suggested alternative but similar proposals;	1						
	suggested several alternatives, each with clear differences. Mark to be dependent on quantity of ideas;	2-4						
	many alternatives, each of original character	5						
5.3 Extent to which these ideas meet the specification. *The candidate's ideas have:*	failed to meet the specification;	0						
	met some of the criteria in the specification;	1-4						
	met all of the criteria in the specification.	5						
5.4 Reasons for selection.	None.	0						
	Low level judgements.	1						
	High level judgements.	2						
6 Graphical Communication of Ideas 6.1 Choice of appropriate techniques. *The candidate's techniques are:*	inappropriate;	0						
	mark to be awarded dependent on extent of effectiveness of the candidate's choice;	1-5						
	completely effective.	6						
6.2 Quality of representation. *The candidate's representation is:*	not recognisable;	0						
	barely recognisable;	1						
	difficult to interpret but conveys information;	2-4						
	competently drawn and conveys increasingly more information;	5-7						
	well defined with a suggestion of flair;	8-9						
	clear, well defined, and showing considerable flair.	10-11						
7 Solution *The candidate's solution:*	was not submitted;	0						
	is obviously not a viable solution but is recognisable as an attempt to define a solution;	1-3						
	is recognisable as a solution but is easily faulted and/or difficult to understand or fails to match the specification;	4-6						
	is incomplete and requires further work but has been presented in a readily understood form;	7-9						
	is substantially free from error and is produced in a clear, descriptive manner which substantially satisfies the specification;	10-13						
	is well conceived, clearly and attractively defined and fully satisfies the specification.	14-17						
8 Evaluation *The candidate's evaluation:*	has not been attempted;	0						
	is irrelevant;	1						
	is relevant but superficial;	2-3						
	represents an honest attempt to appraise his or her work but lacks objectivity and is incomplete;	4-5						
	is complete but lacking in relevance and objectivity;	6-7						
	is complete and largely relevant but lacking in objectivity;	8-10						
	is thorough, objective, relevant and concise; it would provide a useful source of reference for later material.	11-12						
			M	T		M	T	
	Topic Marks	100						
	TOTAL MARK	200						

▲ *Fig. 5 Design and Communication Coursework internal assessment form (NEA)* familiar. However, what is perhaps less familiar is the structure of the paper. This has been dealt with in Chapter 2 and examples are given on pages 31–6.

Paper 2: Design

This is an innovation that has developed over the last twenty years and has formed an integral part of the assessment pattern. Three main approaches have been adopted by GCE and CSE Boards, and it appears that these approaches have been carried through to the GCSE Regional Examination Groups.

The first approach is to present a design problem with sufficient detail to enable a solution to be resolved in the space of approximately two hours without reference to any source material. The second approach, and one which offers much more scope for the candidate, is where a topic is announced weeks before the examination to enable the candidate to prepare information and to do some research to become familiar with the topic. The resultant work can then be taken into the examination room and used as source or reference material to resolve a problem. The problem is, of course, new to the candidate, and as in the first approach has to be resolved observing the codes of conduct of a formal examination.

The third approach links the design element with a practical realisation, as in the East Anglian and London Design and Realisation syllabus. Here the candidate is presented with a number of design problems set by the Group and given a period of three months in which to select a problem, carry out the necessary investigation and complete the work as far as arriving at a 'working drawing'. A copy of the 'working drawing' is sent to the Group and one is retained by each candidate, because some time later they will make what they have designed in a workshop under strictly controlled examination conditions. There is room for modification to ensure that the designs can be made in the time available, which is two sessions totalling five hours. The teacher keeps the examining body informed of any changes.

In this particular syllabus the teacher is also responsible for doing the first stage of marking the realisation which carries a weighting of 15 per cent, thus placing the teacher more and more into the role of an assessor/examiner. With the 40 per cent weighting for coursework, the teacher makes a major contribution to the marking and assessment of students' work.

CONCLUSION

Pupils have already begun the courses that are to be assessed and examined in the summer of 1988. Their teachers, in the main, are planning ahead only as far as the first year, some even less than that, thus highlighting the uncertainties that still exist in their minds. The many and complex problems of inadequate inservice training for teachers and cutbacks in a profession already desperately in need of resources have not helped to provide the ideal set of circumstances for launching a new examination, particularly one that has introduced many radical changes. However, the practice of running two parallel types of examinations, GCE and CSE, was long overdue for change, and to have continued implementing such a system would have been doing a disservice to the younger generation. Thus, whether or not the teaching profession and the examination bodies are fully prepared, GCSE has arrived and the hard work of making it successful has begun.

As for the future of CDT, the prospects have never been better. Throughout its troubled life of trying to gain academic respectability and acceptability with the public at large, the importance of technology in society has never been so great as it is at the present time. The future generations require technologists of all levels, and more important still is the need to provide generations of people who will know how to use the world's resources wisely and to provide an even better world for many generations to follow. In order to achieve these goals the task lies very much in the hands of the teachers. The radical changes that have taken place in the 1980s are only the beginning, and we must not wait another two decades before change emerges again.

SELECTED BIBLIOGRAPHY

GCSE *National Criteria – Craft, Design and Technology*, Department of Education and Science, Welsh Office.

Differentiated Assessment in GCSE – Working Paper No.1, Secondary Examinations Council.

Coursework Assessment in GCSE – Working Paper No.2, Secondary Examinations Council.

Policy and Practice in School-Based Assessment – Working Paper No.3, Secondary Examinations Council.

Craft, Design and Technology: Draft Grade Criteria. Report of Working Party, Secondary Examinations Council.

Craft, Design and Technology: A Guide for Teachers, Secondary Examinations Council/Open University.

Developments in Design Education. John Eggleston, Open Books.

Appendix
EXAMINATION GROUP ADDRESSES

Southern Examining Group
The South-East Regional Examinations Board
Beloe House
2-4 Mount Ephraim Road
Royal Tunbridge Wells
Kent
TN1 1EU

Southern Regional Examinations Board
53 London Road
Southampton
SO9 4YL

The Associated Examining Board
Wellington House
Station Road
Aldershot
Hants
GU11 1BQ

University of Oxford Delegacy of Local Examinations
Ewert Place
Summerton
Oxford
OX2 7BZ

South-Western Examinations Board
23–29 Marsh Street
Bristol
BS1 4BP

Midland Examining Group
East Midland Regional Examinations Board
Robins Wood House
Robins Wood Road
Aspley
Nottingham
NG8 3NR

Oxford and Cambridge Schools Examination Board
Cambridge Office
10 Trumpington Street
Cambridge
CB2 1QB

Oxford Office
Elsfield Way
Oxford
OX2 8EP

University of Cambridge Local Examinations Syndicate
Syndicate Buildings
1 Hills Road
Cambridge
CB1 2EU

The West Midlands Examinations Board
Norfolk House
Smallbrook
Queensway
Birmingham
B5 4NJ

Southern Universities' Joint Board for Examinations
Cotham Road
Bristol
BS6 6DD

London and East Anglian Group
East Anglian Examinations Board
'The Lindens'
Lexden Road
Colchester
CO3 3RL

London Regional Examining Board
Lyon House
104 Wandsworth High Street
London
SW18 4LF

University of London School Examinations Board
Stewart House
32 Russell Square
London
WC1B 5DN

Northern Examination Association
Associated Lancashire Schools Examining Board
12 Harter Street
Manchester
M60 7LH

Joint Matriculation Board
Manchester
M15 6EU

Northern Regional Examinations Board
Wheatfield Road
Westerhope
Newcastle upon Tyne
NE5 5JZ

North West Regional Examinations Board
Orbit House
Albert Street
Eccles
Manchester
M30 0WL

Yorkshire and Humberside Regional Examinations Board
Harrogate Office
31–33 Springfield Avenue
Harrogate
HG1 2HW

Yorkshire and Humberside Regional Examinations Board
Sheffield Office
Scarsdale House
136 Derbyshire Lane
Sheffield
S88 8SE

Northern Ireland Examinations Council
Beechill House
42 Beechill Road
Belfast
BT8 4RS

Welsh Joint Education Committee
245 Western Avenue
Cardiff
CF5 2YX